AF340692

Masked Mycotoxins in Food
Formation, Occurrence and Toxicological Relevance

Issues in Toxicology

Series Editors:
Professor Diana Anderson, *University of Bradford, UK*
Dr Michael Waters, *Integrated Laboratory Systems Inc., N Carolina, USA*
Dr Timothy C. Marrs, *Edentox Associates, Kent, UK*

Advisor to the Board:
Dr Alok Dhawan, *Ahmedabad University, India*

Masked Mycotoxins in Food
Formation, Occurrence and Toxicological Relevance

Edited by

Chiara Dall'Asta
University of Parma, Italy
Email: chiara.dallasta@unipr.it

Franz Berthiller
University of Natural Resources and Life Sciences, Vienna, Austria
Email: franz.berthiller@boku.ac.at

Issues in Toxicology No. 24

Print ISBN: 978-1-84973-972-6
PDF eISBN: 978-1-78262-257-4
ISSN: 1757-7179

A catalogue record for this book is available from the British Library

Published by The Royal Society of Chemistry,
Thomas Graham House, Science Park, Milton Road,
Cambridge CB4 0WF, UK

Registered Charity Number 207890

For further information see our web site at www.rsc.org

Printed in the United Kingdom by CPI Group (UK) Ltd, Croydon, CR0 4YY, UK

Foreword

The safety of food and feed—in particular the presence of food contaminants—has become of increasing concern for consumers, governments and producers. Trace levels of chemical contaminants can originate from natural sources such as mycotoxins, which are secondary metabolites produced by fungi on agricultural commodities in the field and during storage under a wide range of climatic conditions. The occurrence of mycotoxin contamination in various crops is of major concern since it has significant implications for food and feed safety, food security and international trade. The Food and Agriculture Organization (FAO) has estimated that 25% of the world's food crops are affected by mycotoxins, including many basic foodstuffs and animal feeds. In fact, due to the availability of ultra-sensitive high-performance analytical instrumentation, especially in modern liquid chromatography–mass spectrometry (LC-MS), the percentage of samples that have tested positive for mycotoxins in more recent studies is actually much higher. More than 300 mycotoxins have been identified so far with widely different chemical structures and differing modes of action—some target the kidney, liver or the immune system and some are carcinogenic. Common mycotoxins include trichothecenes, such as deoxynivalenol, fumonisins, zearalenone, ochratoxin A and aflatoxins. The potential health risks to animals and humans posed by food- and feed-borne mycotoxin intoxication have been recognised by national and international institutions and organisations such as the European Commission (EC) and its European Food Safety Authority (EFSA), the US Food and Drug Administration (FDA), the World Health Organisation (WHO) and the FAO, which has resulted in improved risk assessment and adopted regulatory limits for major mycotoxin classes and selected individual mycotoxins.

The term "masked mycotoxins", introduced in 1990, has now been established internationally as mycotoxin derivatives that are undetectable by

Issues in Toxicology No. 24
Masked Mycotoxins in Food: Formation, Occurrence and Toxicological Relevance
Edited by Chiara Dall'Asta and Franz Berthiller
© The Royal Society of Chemistry 2016
Published by the Royal Society of Chemistry, www.rsc.org

conventional analytical techniques because their structure has been changed in the plant. In the last decade, masked mycotoxins have become a hot topic in mycotoxin research. To harmonise future scientific wording and subsequent legislation, it has been suggested to use the term "masked mycotoxins" for the fraction of biologically modified mycotoxins that were conjugated by plants.

The role of plant metabolites of mycotoxins is manifold. First and foremost, plants are capable of overcoming or at least diminishing fungal invasion by a variety of mechanisms. The metabolisation of xenobiotics to less deleterious compounds, such as masked mycotoxins, obviously is an important one. Plant breeding efforts have been leading to the selection for those varieties that are more efficient at detoxifying pathogenicity or virulence factors, such as certain mycotoxins. It is therefore not too surprising that a multitude of masked mycotoxins must exist. The tremendous pace in the development of modern analytical equipment and methods has enabled the discovery of many such compounds, in particular during the last couple of years. While there are still plenty of shortcomings to overcome—including the lack of analytical standards, the trueness of analytical results or matrix reference materials—the (analytical) community has taken some important steps in the right direction.

Food safety has to address all compounds with potential negative health effects. Hence, masked mycotoxins along with their parent compounds have to be considered within a sound risk assessment analysis. If and to what extent masked mycotoxins are risk factors for humans and animals is a question that will keep scientists in the field of mycotoxins busy for years to come. The assumption that masked mycotoxins are cleaved during digestion took surprisingly long to prove, showing how complex the topic is. While certain gut microbes that are more abundant in certain animal species are quite capable of cleaving masked mycotoxins, others are not. Only recently has the direct action of masked mycotoxins before cleavage been assessed. Equally importantly, masked mycotoxins might liberate toxins in areas of the body in which mycotoxins normally do not occur.

This is the first book that is exclusively dedicated to the topic of masked mycotoxins and all its facets. It will provide the interested reader with an excellent overview on the topic as well as with detailed insights into the rapidly developing field of these important mycotoxin metabolites produced by plants. In particular, analytical methods, the occurrence of masked mycotoxins, the potential effects of food processing and *in vitro* and *in vivo* toxicity assessment, as well as detoxification strategies for mycotoxins in plant breeding, are discussed in dedicated chapters. *Masked Mycotoxins in Food* enables both the newcomer and the veteran in the field to get a full picture of the current knowledge on masked mycotoxins.

Enjoy reading this book!

Rudolf Krska
University of Natural Resources and Life Sciences, Vienna, Austria

Contents

Issues in Toxicology No. 24
Masked Mycotoxins in Food: Formation, Occurrence and Toxicological Relevance
Edited by Chiara Dall'Asta and Franz Berthiller
© The Royal Society of Chemistry 2016
Published by the Royal Society of Chemistry, www.rsc.org

List of Abbreviations

3-Ac-DON	3-acetyl-deoxynivalenol
4-OH-OTA	4-hydroxy-ochratoxin A
15-Ac-DON	15-acetyl-deoxynivalenol
ABC	ATP-binding cassette
ADME	absorption, distribution, metabolism and excretion
AFB_1	aflatoxin B_1
AFB_2	aflatoxin B_2
AFG_1	aflatoxin G_1
AFG_2	aflatoxin G_2
AFM_1	aflatoxin M_1
AFP_1	aflatoxin P_1
AFQ_1	aflatoxin Q_1
AF	aflatoxin
ATP	adenosine triphosphate
AUC	area under the curve
BSA	bovine serum albumin
b.w.	body weight
CMO	carboxymethyloxime
CONTAM	Panel on Contaminants in the Food Chain
CR	cross-reactivity
CYP	cytochrome P450
DAS	diacetoxyscirpenol
DAS-3-Glc	diacetoxyscirpenol-3-glucoside
DDA	data-dependent analysis
DNA	deoxyribonucleic acid
DOM-1	de-epoxy deoxynivalenol
DOM-1-S	de-epoxy deoxynivalenol-3-sulphate
DON	deoxynivalenol

Issues in Toxicology No. 24
Masked Mycotoxins in Food: Formation, Occurrence and Toxicological Relevance
Edited by Chiara Dall'Asta and Franz Berthiller
© The Royal Society of Chemistry 2016
Published by the Royal Society of Chemistry, www.rsc.org

DON-3-Glc	deoxynivalenol-3-glucoside
DON-3-GlcA	deoxynivalenol-3-glucuronide
DON-7-GlcA	deoxynivalenol-7-glucuronide
DON-15-GlcA	deoxynivalenol-15-glucuronide
DON-GSH	deoxynivalenol-glutathione conjugate
DON-S	deoxynivalenol-3-sulphate
EA	ergot alkaloid
EFSA	European Food Safety Authority
ELISA	enzyme-linked immunosorbent assay
ENNB	enniatin B
$ENNB_1$	enniatin B_1
ENN	enniatin
F	oral bioavailability
F_A	transport across the intestinal epithelium
FAO	Food and Agriculture Organization of the United Nations
F_B	bioaccessibility
FB_1	fumonisin B_1
FB_2	fumonisin B_2
FBs	fumonisins
F_H	first-pass effect
FHB	*Fusarium* head blight
FLD	fluorescence detection
FPIA	fluorescence polarisation immunoassay
FTICR	Fourier transform ion cyclotron resonance
FUSX-3-Glc	fusarenon-X-3-glucoside
FWHM	full width of the peak at half its maximum
GIT	gastrointestinal tract
HFB1	hydrolysed fumonisin B_1
HPLC	high-performance liquid chromatography
HRMS	high-resolution mass spectrometry
HT2	HT-2 toxin
HT2-3-diGlc	HT-2 toxin-3-diglucoside
HT2-3-Glc	HT-2 toxin-3-glucoside
HT2-4-Glc	HT-2 toxin-4-glucoside
IAC	immunoaffinity column
IC_{50}	half maximal inhibitory concentration
IDA	information data analysis
iSPR	imaging surface plasmon resonance sensor
JECFA	Joint FAO/WHO Expert Committee on Food Additives
LC-MS/MS	liquid chromatography–tandem mass spectrometry
LD_{50}	median lethal dose
LDA	linear discriminant analysis
LFD	lateral flow device
LOAEL	lowest observed adverse effect level
$\log D$	distribution coefficient

log$P$	octanol/water partition coefficient
lysil-FB$_1$	mono- and di-lysine derivatives of fumonisin B$_1$
m/z	mass-to-charge ratio
mAb	monoclonal antibody
MAS	monoacetoxyscirpenol
MAS-3-Glc	monoacetoxyscirpenol-3-glucoside
MON	moniliformin
mRNA	messenger ribonucleic acid
MS	mass spectrometry
NAC	N-acetylcysteine
NCM-FB$_1$	N-(carboxymethyl) fumonisin B$_1$
NDF-FB$_1$	N-deoxy-fructosyl fumonisin B$_1$
NDF-HFB$_1$	N-deoxy-fructosyl hydrolysed fumonisin B$_1$
NEO	neosolaniol
NEO-3-Glc	neosolaniol-3-glucoside
NIV	nivalenol
NIV-3-Glc	nivalenol-3-glucoside
NMR	nuclear magnetic resonance
NOAEL	no observed adverse effect level
OTA	ochratoxin A
OTα	ochratoxin α
PAT	patulin
PCA	principal component analysis
pHFB$_1$	partially hydrolysed fumonsin B$_1$
PMTDI	provisional maximum tolerable daily intake
ppm	parts per million
qPCR	quantitative real-time polymerase chain reaction
QTOF	quadrupole-time-of-flight
RR	relative response
Sa	sphinganine
Sa/So	sphinganine-to-sphingosine ratio
SDS	sodium dodecyl sulphate
SIL	stable isotopic labelling
So	sphingosine
SPR	surface plasmon resonance sensor
T2	T-2 toxin
T2-3-diGlc	T-2 toxin-3-diglucoside
T2-3-Glc	T-2 toxin-3-glucoside
TOF	time-of-flight
UDP	uridine diphosphate
UHPLC	ultra-high-performance liquid chromatography
ZEN	zearalenone
ZEN-14-Glc	zearalenone-14-glucoside
ZEN-14-S	zearalenone-14-sulfate
ZEN-16-Glc	zearalenone-16-glucoside
αZAL	α-zearalanol

αZEL   α-zearalenol
αZEL-14-Glc   α-zearalenol-14-glucoside
βZAL   β-zearalanol
βZEL   β-zearalenol
βZEL-14-Glc   β-zearalenol-14-glucoside

αZEL   α-zearalenol
αZEL-14-Glc   α-zearalenol-14-glucoside
βZAL   β-zearalanol
βZEL   β-zearalenol
βZEL-14-Glc   β-zearalenol-14-glucoside

Introduction to Masked Mycotoxins

FRANZ BERTHILLER,[*,a] CHRIS M. MARAGOS[b] AND CHIARA DALL'ASTA[c]

[a] Center for Analytical Chemistry, Department for Agrobiotechnolgy (IFA-Tulln), University of Natural Resources and Life Sciences, Vienna (BOKU), Konrad Lorenz Straße 20, 3430 Tulln, Austria;
[b] USDA–ARS–NCAUR, 1815 N. University Street, Peoria, IL 61604, USA;
[c] Department of Food Science, University of Parma, Parco Area Scienze 17/A, 43124 Parma, Italy
*Email: franz.berthiller@boku.ac.at

1.1 Mycotoxins

Given suitable water activities, moulds can infect almost every agricultural commodity (*e.g.* cereals, nuts, fruits, *etc.*) during plant growth and/or after harvest. A variety of these fungi, in particular *Aspergillus* spp., *Penicillium* spp. or *Fusarium* spp., are capable of producing mycotoxins. These substances are low-molecular-weight fungal secondary metabolites that are able to accumulate in food or feed in toxicologically relevant concentrations.[1] About 20 years ago, the Food and Agriculture Organization (FAO) of the United Nations estimated that 25% of the world's food crops were significantly contaminated with mycotoxins, leading to an annual loss in the range of 1000 million tons.[2] Recent studies suggest that the percentage of contaminated cereals is much higher: 72%.[3] The difference may be due, in part, to what levels are regarded as contamination, as well as improvements in monitoring.

Issues in Toxicology No. 24
Masked Mycotoxins in Food: Formation, Occurrence and Toxicological Relevance
Edited by Chiara Dall'Asta and Franz Berthiller

Published by the Royal Society of Chemistry, www.rsc.org

The potential of mycotoxins to cause harm to human health through dietary exposure has led authorities world-wide to highly regulate these food contaminants. By the end of 2003, about 100 countries had regulations for maximum levels of mycotoxins in various food and feedstuffs.[4] For instance, in the European Union, maximum levels have been set for the mycotoxins aflatoxin B_1 (AFB$_1$), B_2 (AFB$_2$), G_1 (AFG$_1$) and G_2 (AFG$_2$), deoxynivalenol (DON), fumonisin B_1 (FB$_1$) and B_2 (FB$_2$), ochratoxin A (OTA) and patulin (PAT), as well as for zearalenone (ZEN) in foodstuffs.[5] Furthermore, indicative levels have been recommended for T-2 and HT-2 toxins (T2 and HT2) in a variety of cereals and products thereof.[6]

1.2   Masked Mycotoxins

Unaltered mycotoxins are not the only source of mycotoxin exposure for consumers. Plants protect themselves from xenobiotic compounds like mycotoxins by converting them into more polar metabolites (Figure 1.1), which are transported into vacuoles for further storage or are conjugated to biopolymers such as cell wall components.[7,8] Mycotoxins, which are in contact with highly metabolically active plants in the field, are especially prone to being metabolised. As *Fusarium* infection usually occurs in the field (in contrast to *Aspergillus* or *Penicillium* infections), the *Fusarium* mycotoxins DON, ZEN, FB$_1$, T2, HT2 and nivalenol (NIV) are the most prominent targets for conjugation.

1.2.1   Terminology

The formed substances are often referred to as "masked mycotoxins". Unfortunately, within the literature there are sufficient inconsistencies in

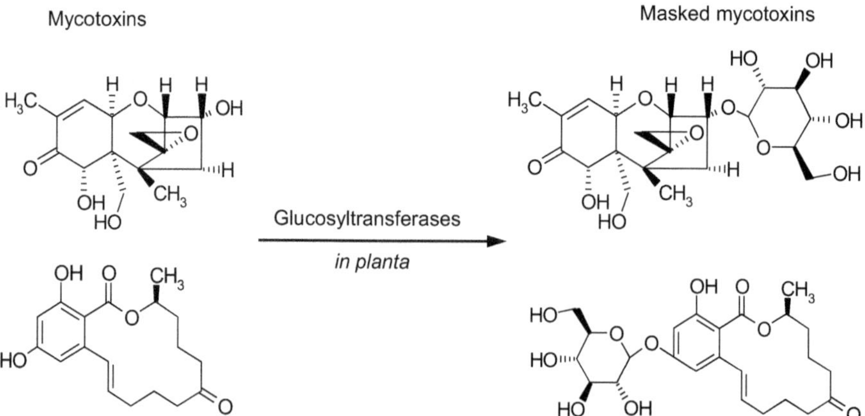

Figure 1.1   Conversion of the *Fusarium* mycotoxins deoxynivalenol (top left) and zearalenone (lower left) into deoxynivalenol-3-glucoside (top right) and zearalenone-14-glucoside (lower right) by plants.

terms that a brief description of how they will be used in this book is needed. The terms "masked", "hidden", "conjugated" and "bound" are frequently used in the literature. Of these, "masked" is the most popular. The term was originally intended to distinguish substances that were the targets during routine analysis from those that, while not targeted analytes, might contribute to the mycotoxin content.[9] In this context, mycotoxin derivatives can arise through a number of mechanisms. They may be precursors, metabolites or degradation products of the "parent" (or free) form of the mycotoxin, or they may have been formed abiotically through chemical reaction of the parent toxin with the matrix (*e.g.* through food processing).[8,10] Unfortunately, with this definition, once a masked mycotoxin is a routine target of analysis, it is no longer truly a masked mycotoxin. Despite this contradiction, the definition is widely used because it has merit as an inclusive term for a wide variety of materials that have traditionally been overlooked and that might contribute to toxicity. Masked mycotoxins can be further classified according to how the masked form relates to the parent form; that is, whether the masked form exists as a covalent derivative of the parent toxin or a non-covalent association between the parent toxin and a matrix component.[7,11] Within the literature, some references describe covalently linked forms as "conjugated", while the non-covalently (*e.g.* associated) forms have been termed "bound". The latter term is intended to indicate that the parent toxin might be extractable from the non-covalent complex following chemical or enzymatic treatment.[12] This distinction is important, but conjugates, because they contain covalent linkages, are also bound forms of the parent toxin. Perhaps the best description of "hidden" mycotoxins was provided by Dall'Asta *et al.*,[11] where they were classified as non-covalent, associative interactions between a mycotoxin and matrix macroconstituents.

The most recent and comprehensive definition was worked out within the scope of the Committee for Contaminants and other Undesirable Substances in the Food Chain of the German Federal Institute for Risk Assessment (BfR).[13] In this systematic definition, distinctions were made between free mycotoxins, matrix-associated mycotoxins and "modified mycotoxins". Modified mycotoxins were further classified into those that are biologically or chemically modified. Regarding biological modifications, these can be achieved, for example, by plants, animals, fungi or other means. Using the BfR proposal, only plant metabolites of mycotoxins would still be termed masked mycotoxins. This definition was taken up by the Panel on Contaminants in the Food Chain (CONTAM) of the European Food Safety Authority (EFSA) in their recent scientific opinion on modified mycotoxins,[14] as well by the authors of this book.

1.2.2 Historical Perspective

The issue of masked mycotoxins began attracting scientific interest after several mysterious cases of mycotoxicosis during the mid-1980s, in which symptoms in affected animals did not correlate with the low mycotoxin content detected in their feed. Around the same period, the metabolic bio-transformation of DON to less toxic derivatives *in planta* was for the first time hypothesised to occur in field corn inoculated with *Fusarium graminearum*[15] and in naturally infected winter wheat.[16] It was shown that callus cultures of the *Fusarium* head blight-resistant wheat cultivar Frontana con-verted more ^{14}C-labelled DON into uncharacterised products than callus derived from the susceptible wheat cultivar Casavant.[17] Later, the major soluble DON metabolite of plants, deoxynivalenol-3-β-D-glucoside (DON-3-Glc; Figure 1.1), was isolated from DON-treated maize cell suspension cultures.[18] Another decade later, the substance was shown to arise after treatment of *Arabidopsis thaliana* with DON,[19] until it was for the first time also found in naturally contaminated maize and wheat.[20] It has been shown that, after treatment of plants with DON, wheat produces DON-3-Glc to de-toxify this *Fusarium graminearum* virulence factor.[21] A survey demonstrated that DON-3-Glc concentrations can exceed 1000 μg kg^{-1} in naturally con-taminated wheat and can reach over 70% of the molar DON concentration in maize.[22] DON-3-Glc can also be found in naturally contaminated barley,[23] beer made thereof,[24] breakfast cereals and snacks.[25] The relative proportion of DON-3-Glc to DON in cereals can vary considerably, but on average is in the range of 20%.[26]

ZEN, a *Fusarium* mycotoxin with high oestrogenic activity, was the next piece to find its place in the masked mycotoxin puzzle, when wheat and maize cell cultures were found to be capable of transforming ZEN into zearalenone-14-β-D-glucopyranoside (ZEN-14-Glc) and other metabolites.[27] The same study indicated that about 90% of the added radiolabelled ^{14}C-ZEN was recovered in soluble form after 3 days. A later paper by the same group reported that, 12 days after treatment, more than 50% of the radio-activity was found in the non-extractable residues.[28] The bioavailability of these bound forms has yet to be determined. Liquid chromatography–tandem mass spectrometry (LC-MS/MS) studies have proven that the model plant *Arabidopsis thaliana* can rapidly transform ZEN into an array of 17 different compounds (Figure 1.2), including glucosides, malonylgluco-sides, diglucosides and pentosylhexosides of ZEN and their phase I metab-olites α-zearalenol (αZEL) and β-zearalenol (βZEL).[29]

Fumonisin conjugates were long believed to occur only after food pro-cessing.[31] However, it was shown that bound fumonisins could also be found in unprocessed maize.[32] The exact chemical nature of these naturally occurring hidden forms is still unknown. Most likely, non-covalent inter-actions with, for example, starch or proteins occur, rendering the fumoni-sins difficult to extract.[11]

Generally, very little is known regarding bound mycotoxins. The IUPAC has proposed the following general definition for bound residues: "A xenobiotic bound residue is a residue which is associated with one or more classes of endogenous macromolecules. It cannot be disassociated from the natural macromolecule using exhaustive extraction or digestion without

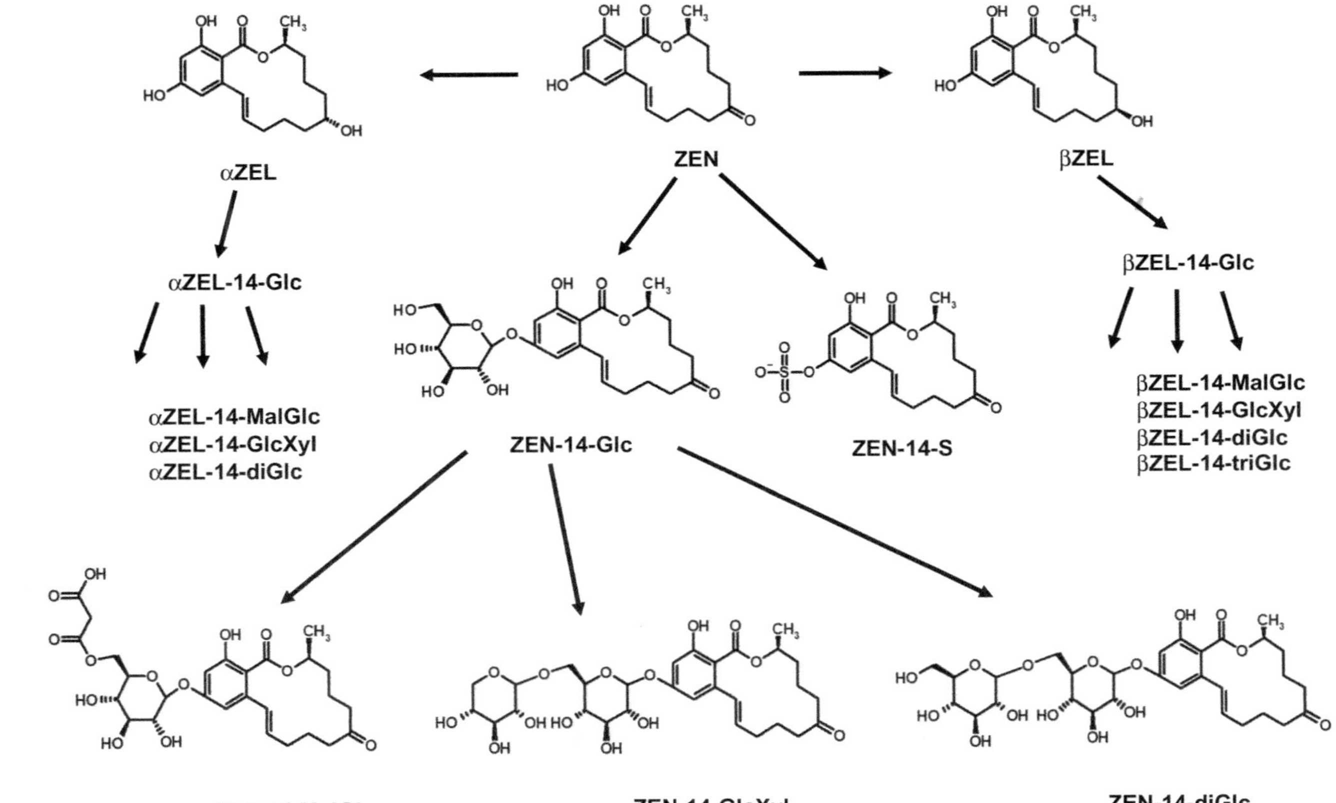

Figure 1.2 Metabolism of ZEN in *Arabidopsis thaliana*. (Modified from Berthiller *et al.*[30])

significantly changing the nature of either the exocon or the associated endogenous macromolecules."[33] Depending on the type of linkage to proteins, starch, pectins, hemicellulose, cellulose and lignin, it is conceivable that at least a part of bound mycotoxins could become bioavailable again in the digestive tract of humans and animals. Bound residues are either covalent or non-covalent. Bound residues are usually quantified as the difference of radioactivity compared to the soluble fraction after treatment of plants with the radionuclide-labelled xenobiotics of interest. Depending on the chemical nature of the xenobiotic, very different incorporation rates have been found. Pesticides, for example, showed incorporation rates from just a few to up to 90% of the compound applied to the plant.[34]

1.2.3 Recent Developments

Additional DON conjugates have been identified recently. Oligoglycosides of DON, namely di-, tri- and tetra-glucosides, have been found in beer.[35] The formation of a DON–glutathione conjugate has been shown *in vitro*,[36] and the occurrence of this compound in cereals was confirmed 3 years later. To do so, a liquid chromatography–high-resolution mass spectrometry (LC-HRMS)-based approach using *in vivo* stable isotopic labelling, combined with a newly developed software tool[37] to extract biological features originating from true metabolites, was employed.[38] Flowering wheat ears were inoculated with a mixture of DON and ^{13}C-labelled DON. In addition to DON-3-Glc, DON–glutathione and its processing products DON-S-cysteine and DON-S-cysteinyl–glycine, as well as DON-malonyl-glucoside, were found. In a continuation of this work, tentative annotation of the remaining biotransformation products was carried out, additionally identifying DON-hexitol (*e.g.* mannitol), DON-di-hexoside (*e.g.* glucose), 15-acetyl-DON-3-β-D-glucoside and a DON–glutathione derivative lacking two protons.[39] Most recently, also DON-3-sulphate and DON-15-sulphate have been identified in *Fusarium graminearum*-inoculated or DON-treated wheat.[40]

Several new masked mycotoxins have also been described during the last few years. Both nivalenol-3-glucoside (NIV-3-Glc) and fusarenon-X-glucoside (FUSX-3-Glc) were found in wheat grain using high-resolution mass spectrometry.[41] Furthermore, several T2 (T2-Glc) and HT2 glucosides (HT2-Glc) were detected in contaminated wheat and oats by LC-MS/MS[42] or LC-HRMS,[43] respectively. The di-glucosides of both T2 and HT2 have also been reported.[44] Glucosides of other type A trichothecenes, namely neosolaniol-glucoside (NEO-Glc) and diacetoxyscirpenol-glucoside (DAS-Glc), were found in maize powder.[45] Most recently, a new ZEN-glucoside, which was isolated from ZEN-treated barley, was found and determined to be ZEN-16-β-D-glucopyranoside (ZEN-16-Glc).[46]

Analytical methods for the determination of masked mycotoxins have been summarised.[7,47] Moreover, a comprehensive review on the occurrence of masked mycotoxins in food and analytical aspects for their determination, toxicology and impact on stakeholders were published.[8] In brief, there exist at least three analytical strategies to determine masked mycotoxins in food. The first of these are dedicated LC-MS/MS-based methods, which can quantify masked mycotoxins along with their parent forms. Secondly, masked mycotoxins can be detected by immunochemical methods, provided there is cross-reactivity of antibodies towards them. Both LC-MS/MS and immunoassay techniques are strengthened with the availability of appropriate analytical standards of the masked forms. Finally, masked mycotoxins may be hydrolysed to their parent mycotoxins using enzymes or harsh acidic or alkaline conditions. A sum parameter (parent and masked mycotoxins) is derived with the latter two approaches and the masked fraction might be calculated by subtracting the concentration of the parent mycotoxin determined by conventional techniques. Those indirect techniques should be used carefully, however. A recent paper describes the inability of previously published works based on acidic hydrolysis for the determination of masked DON in cereals, none of which were able to liberate DON from its major metabolite DON-3-Glc.[48] Enzymatic cleavage by certain β-glucosidases seems to be far more promising for liberating the parent toxins from DON-3-Glc, NIV-3-Glc or HT2-Glc in cereal matrices.[49] The purified recombinant enzyme from *Bifidobacterium adolescentis* works rapidly under the given conditions, allowing complete cleavage of DON-3-Glc in cereal extracts within 10 minutes of incubation. The interested reader is referred to Chapters 3 and 4 for far more detailed information on the determination of masked mycotoxins using immunoanalytical and mass spectrometric techniques.

1.2.4 Toxicity of Masked Mycotoxins

In general, intact masked mycotoxins are less potent relatives to their unmodified forms.[8] This is easily understandable, taking into account the severe modifications of the toxins due to conjugation and the fact that masked mycotoxins arise during detoxification reactions of plants. For instance, compared to DON, DON-3-Glc barely binds to the ribosome, resulting in a highly diminished inhibition of protein synthesis—the major mode of action for all trichothecenes.[19] Similarly, ZEN-14-Glc can barely bind to the oestrogen receptor, resulting in a far reduced oestrogenicity compared to its parent toxin ZEN.[50] However, it is assumed that masked mycotoxins can be "reactivated" during mammalian digestion by cleavage of the polar group and liberation of the native toxin. Perhaps because toxicity testing can be very time consuming, the toxic effects caused by masked mycotoxins are only beginning to appear in the literature. Even so, a wealth of information has been gained on DON-3-Glc, especially within the past 3 years. In 2011, the Joint FAO/WHO Expert Committee on Food Additives (JECFA) emphasised the occurrence of DON-3-Glc in cereals and beers, which might contribute to systemic exposure to DON.[51] Besides recommending additional studies to collect occurrence data as well as to investigate the effects of processing on DON-3-Glc, the JECFA also asked for absorption, distribution, metabolism and excretion (ADME) studies on this substance. In the same year, the hydrolytic fate of DON-3-Glc during digestion was assessed using several *in vitro* assays.[52] DON-3-Glc proved to be stable towards hydrochloric acid and human enzymes, but several lactic acid bacteria showed the ability to partially release DON. Two independent studies verified the results, showing that the release of DON also occurs after incubating DON-3-Glc with human faeces.[53,54] Nevertheless, *in vivo* ADME studies were necessary to assess the potential health risk of DON-3-Glc. The fate of orally administered DON-3-Glc was determined in rats and piglets.[55,56] It was concluded that DON-3-Glc was less bioavailable than DON, but was almost completely hydrolysed in both species. The cleavage took place mostly in the hindgut, where absorption is lower than in the small intestine. Due to differences in anatomy and gut microbiota, the metabolism was species dependent. In addition, the state of digestion and individual differences in gut microbiota can cause differences in the amount of DON released. Again, readers interested in the toxicological aspects of masked mycotoxins are referred to Chapters 6 and 7, where *in vitro* and *in vivo* experiments are discussed in far more detail.

1.3   Conclusion

Rapid developments in analytical chemistry, in particular the rise of mass spectrometry, have been instrumental in the identification and detection of a variety of formerly unknown plant metabolites of mycotoxins. Following detection, the next logical step is to elucidate the prevalence and levels of masked mycotoxins in foodstuffs. This task is currently still hampered by the (non-)availability of reference standards for the community. Several research groups are actively working on the synthesis or isolation of masked mycotoxins, so this bottleneck should be overcome soon. Accurate risk assessment of masked mycotoxins also requires assessment of their toxicity. In many cases, the precursor toxins are (partly) liberated by gut bacteria. Still, as each compound has to be verified individually, this challenging task will need more research to sufficiently answer the questions raised for years to come.

References

1. J. W. Bennett and M. Klich, Mycotoxins, *Clin. Microbiol. Rev.*, 2003, **16**, 497–516.
2. J. E. Smith, G. L. Solomons, C. W. Lewis and J. G. Anderson, Mycotoxins in human nutrition and health. European Commission CG XII, 1994.
3. E. Streit, K. Naehrer, I. Rodrigues and G. Schatzmayr, Mycotoxin occurrence in feed and feed raw materials worldwide: long-term analysis with special focus on Europe and Asia, *J. Sci. Food Agric.*, 2013, **93**, 2892–2899.
4. H. P. van Egmond, R. C. Schothorst and M. A. Jonker, Regulations relating to mycotoxins in food: Perspectives in a global and European context, *Anal. Bioanal. Chem.*, 2007, **389**, 147–157.

5. European Commission Regulation (EC) No 1881/2006 setting maximum levels for certain contaminants in foodstuff – consolidated version 03.12.2012, http://eur-lex.europa.eu/LexUriServ/LexUriServ.do?uri=CONSLEG:2006R1881:20121203:EN:PDF accessed March 30, 2015.

6. European Commission Recommendation (EU) No 2013/165 of 27 March 2013 on the presence of T-2 and HT-2 toxin in cereals and cereal products, *Off. J. Eur. Union*, 2013, **L 91**, 12–15.

7. F. Berthiller, R. Schuhmacher, G. Adam and R. Krska, Formation, determination and significance of masked and other conjugated mycotoxins, *Anal. Bioanal. Chem.*, 2009, **395**, 1243–1252.

8. F. Berthiller, C. Crews, C. Dall'Asta, S. D. Saeger, G. Haesaert, P. Karlovsky, I. P. Oswald, W. Seefelder, G. Speijers and J. Stroka, Masked mycotoxins: A review, *Mol. Nutr. Food Res.*, 2013, **57**, 165–186.

9. M. Gareis, J. Bauer, J. Thiem, G. Plank, S. Grabley and B. Gedek, Cleavage of zearalenone-glycoside, a "masked" mycotoxin, during digestion in swine, *J. Vet. Med., Ser. B*, 1990, **37**, 236–240.

10. S. De Saeger and H. P. van Egmond, Special issue: Masked mycotoxins, *World Mycotoxin J*, 2012, **5**, 203–206.

11. C. Dall'Asta, M. Mangia, F. Berthiller, A. Molinelli, M. Sulyok, R. Schuhmacher, R. Krska, G. Galaverna, A. Dossena and R. Marchelli, Difficulties in fumonisin determination: the issue of bound fumonisins, *Anal. Bioanal. Chem.*, 2009, **395**, 1335–1345.

12. I. Y. Goryacheva and S. De Saeger, Immunochemical detection of masked mycotoxins: A short review, *World Mycotoxin J.*, 2012, **5**, 281–287.

13. M. Rychlik, H.-U. Humpf, D. Marko, S. Dänicke, A. Mally, F. Berthiller, H. Klaffke and N. Lorenz, Proposal of a comprehensive definition of modified and other forms of mycotoxins including "masked" mycotoxins, *Mycotox. Res.*, 2014, **30**, 197–205.

14. EFSA Panel on Contaminants in the Food Chain, Scientific Opinion on the risks for human and animal health related to the presence of modified forms of certain mycotoxins in food and feed, *EFSA J.*, 2014, **12**, 3916.

15. J. D. Miller, J. C. Young and H. L. Trenholm, Fusarium toxins in field corn. I. Time course of fungal growth and production of deoxynivalenol and other mycotoxins, *Can. J. Bot.*, 1983, **61**, 3080–3087.

16. P. M. Scott, K. Nelson, S. R. Kanhere, K. F. Karpinski, S. Hayward, G. A. Neish and A. H. Teich, Decline in deoxynivalenol concentrations in 1983 Ontario winter wheat before harvest, *Appl. Environ. Microbiol.*, 1984, **48**, 884–886.

17. J. D. Miller and P. G. Arnison, Degradation of deoxynivalenol by suspension cultures of the Fusarium head blight resistant wheat cultivar Frontana, *Can. J. Plant Pathol.*, 1986, **8**, 147–150.

18. N. Sewald, J. Lepschy von Gleissenthall, M. Schuster, G. Müller and R. T. Aplin, Structure elucidation of a plant metabolite of 4-desoxynivalenol, *Tetrahedron*, 1992, **3**, 953–960.

19. B. Poppenberger, F. Berthiller, D. Lucyshyn, T. Sieberer, R. Schuhmacher, R. Krska, K. Kuchler, J. Glössl, C. Luschnig and G. Adam, Detoxification of the Fusarium mycotoxin deoxynivalenol by a UDP-glucosyltransferase from *Arabidopsis thaliana*, *J. Biol. Chem.*, 2003, **278**, 47905–47914.

20. F. Berthiller, C. Dall'Asta, R. Schuhmacher, M. Lemmens, G. Adam and R. Krska, Masked Mycotoxins: Determination of a Deoxynivalenol Glucoside in Artificially and Naturally Contaminated Wheat by LC-MS/MS, *J. Agric. Food Chem.*, 2005, **53**, 3421–3425.

21. M. Lemmens, U. Scholz, F. Berthiller, A. Koutnik, C. Dall'Asta, R. Schuhmacher, G. Adam, A. Mesterhazy, R. Krska, H. Buerstmayr and P. Ruckenbauer, A major QTL for Fusarium head blight resistance in wheat is correlated with the ability to detoxify the mycotoxin deoxynivalenol, *Mol. Plant-Microbe Interact.*, 2005, **18**, 1318–1324.

22. F. Berthiller, R. Corradini, C. Dall'Asta, R. Marchelli, M. Sulyok, R. Krska, G. Adam and R. Schuhmacher, Occurrence of deoxynivalenol and its 3-β-D-glucoside in wheat and maize, *Food Addit. Contamin.*, A, 2009, **26**, 507–511.

23. K. Lancova, J. Hajslova, J. Poustka, A. Krplova, M. Zachariasova, P. Dostalek and L. Sachambula, Transfer of Fusarium mycotoxins and 'masked' deoxynivalenol (deoxynivalenol-3-glucoside) from field barley through malt to beer, *Food Addit. Contam.*, 2008, **25**, 732–744.

24. M. Kostelanska, J. Hajslova, M. Zachariasova, A. Malachová, K. Kalachova, J. Poustka, J. Fiala, P. M. Scott, F. Berthiller and R. Krska, Occurrence of deoxynivalenol and its major conjugate, deoxynivalenol-3-glucoside, in beer and some brewing intermediates, *J. Agric. Food Chem.*, 2009, **57**, 3187–3194.

25. A. Malachová, Z. Dzuman, Z. Veprikova, M. Vaclavikova, M. Zachariasova and J. Hajslova, Deoxynivalenol, Deoxynivalenol-3-glucoside, and Enniatins: The Major Mycotoxins Found in Cereal-Based Products on the Czech Market, *J. Agric. Food Chem.*, 2011, **59**, 12990–12997.

26. A. Desmarchelier and W. Seefelder, Survey of deoxynivalenol and deoxynivalenol-3-glucoside in cereal-based products by liquid chromatography electrospray ionization tandem mass spectrometry, *World Mycotoxin J*, 2011, **4**, 29–35.

27. G. Engelhardt, G. Zill, B. Wohner and P. R. Wallnöfer, Transformation of the Fusarium mycotoxin zearalenone in maize cell suspension cultures, *Naturwissenschaften*, 1988, **75**, 309–310.

28. G. Engelhardt, M. Ruhland and P. R. Wallnöfer, Metabolism of mycotoxins in plants, *Adv. Food Sci.*, 1999, **21**, 71–78.

29. F. Berthiller, U. Werner, M. Sulyok, R. Krska, M. T. Hauser and R. Schuhmacher, Liquid chromatography coupled to tandem mass spectrometry (LC-MS/MS) determination of phase II metabolites of the mycotoxin zearalenone in the model plant Arabidopsis thaliana, *Food Addit. Contamin.*, 2006, **23**, 1194–1200.

30. F. Berthiller, M. Lemmens, U. Werner, R. Krska, M. T. Hauser, G. Adam and R. Schuhmacher, Short review: Metabolism of theFusarium

mycotoxins deoxynivalenol and zearalenone in plants, *Mycotox. Res.*, 2007, **23**, 68–72.

31. H.-U. Humpf and K. A. Voss, Effects of thermal food processing on the chemical structure and toxicity of fumonisin mycotoxins, *Mol. Nutr. Food Res.*, 2004, **48**, 255–269.

32. C. Dall'Asta, G. Galaverna, G. Aureli, A. Dossena and R. Marchelli, A LC/MS/MS method for the simultaneous quantification of free and masked fumonisins in maize and maize-based products, *World Mycotoxin J.*, 2008, **1**, 237–246.

33. M. W. Skidmore, G. D. Paulson, H. A. Kuiper, B. Ohlin and S. Reynolds, Bound xenobiotic residues in food commodities of plant and animal origin, *Pure Appl. Chem.*, 1998, **70**, 1423–1447.

34. H. Sandermann, Bound and unextractable pesticidal plant residues: chemical characterization and consumer exposure, *Pest Manage. Sci.*, 2004, **60**, 613–623.

35. M. Zachariasova, M. Vaclavikova, O. Lacina, L. Vaclavik and J. Hajslova, New "Masked" Fusarium Toxins Occurring in Malt, Beer, and Breadstuff, *J. Agric. Food Chem.*, 2012, **60**, 9280–9291.

36. S. A. Gardiner, J. Boddu, F. Berthiller, C. Hametner, R. M. Stupar, G. Adam and G. J. Muehlbauer, Transcriptome analysis of the barley-deoxynivalenol interaction: evidence for a role of glutathione in deoxynivalenol detoxification, *Mol. Plant-Microbe Interact.*, 2010, **23**, 962–976.

37. C. Büschl, B. Kluger, F. Berthiller, G. Lirk, S. Winkler, R. Krska and R. Schuhmacher, MetExtract: A new software tool for the automated comprehensive extraction of metabolite-derived LC/MS signals in metabolomics research, *Bioinformatics*, 2012, **28**, 736–738.

38. B. Kluger, C. Bueschl, M. Lemmens, F. Berthiller, G. Häubl, G. Jaunecker, G. Adam, R. Krska and R. Schuhmacher, Stable isotopic labelling-assisted untargeted metabolic profiling reveals novel conjugates of the mycotoxin deoxynivalenol in wheat, *Anal. Bioanal. Chem.*, 2013, **405**, 5031–5036.

39. B. Kluger, C. Bueschl, M. Lemmens, H. Michlmayr, A. Malachová, A. Koutnik, I. aloku, F. Berthiller, G. Adam, R. Krska and R. Schuhmacher, Biotransformation of the mycotoxin deoxynivalenol in Fusarium resistant and susceptible near isogenic wheat lines, *PLoS One*, 2015, **10**, e0119656.

40. B. Warth, P. Fruhmann, G. Wiesenberger, B. Kluger, B. Sarkanj, M. Lemmens, C. Hametner, J. Fröhlich, G. Adam, R. Krska and R. Schuhmacher, Deoxynivalenol-sulfates: identification and quantification of novel conjugated (masked) mycotoxins in wheat, *Anal. Bioanal. Chem.*, 2015, **407**, 1033–1039.

41. H. Nakagawa, K. Ohmichi, S. Sakamoto, Y. Sago, M. Kushiro, H. Nagashima, M. Yoshida and T. Nakajima, Detection of a new Fusarium masked mycotoxin in wheat grain by high-resolution LC-Orbitrap MS, *Food Addit. Contam., A*, 2011, **28**, 1447–1456.

42. M. Busman, S. M. Poling and C. M. Maragos, Observation of T-2 Toxin and HT-2 Toxin Glucosides from *Fusarium sporotrichioides* by Liquid Chromatography Coupled to Tandem Mass Spectrometry (LC-MS/MS), *Toxins*, 2011, **3**, 1554–1568.
43. V. M. T. Lattanzio, A. Visconti, M. Haidukowski and M. Pascale, Identification and characterization of new Fusarium masked myco-toxins, T2 and HT2 glycosyl derivatives, in naturally contaminated wheat and oats by liquid chromatography-high-resolution mass spectrometry, *J. Mass Spectrom.*, 2012, **47**, 466–475.
44. H. Nakagawa, S. Sakamoto, Y. Sago and H. Nagashima, Detection of type A trichothecene di-glucosides produced in corn by high-resolution liquid chromatography-Orbitrap mass spectrometry, *Toxins*, 2013, **5**, 590–604.
45. H. Nakagawa, S. Sakamoto, Y. Sago, M. Kushiro and H. Nagashima, Detection of masked mycotoxins derived from type A trichothecenes in corn by high-resolution LC-Orbitrap mass spectrometer, *Food Addit. Contam., A*, 2013, **30**, 1407–1414.
46. M. P. Kovalsky Paris, W. Schweiger, C. Hametner, R. Stuckler, G. J. Muehlbauer, E. Varga, R. Krska, F. Berthiller and G. Adam, Zearalenone-16-O-glucoside: a new masked mycotoxin, *J. Agric. Food Chem.*, 2014, **62**, 1181–1189.
47. M. Cirlini, C. Dall'Asta and G. Galaverna, Hyphenated chromatographic techniques for structural characterization and determination of masked mycotoxins, *J. Chromatogr. A*, 2012, **1255**, 145–152.
48. A. Malachová, L. Štočková, A. Wakker, E. Varga, R. Krska, H. Michlmayr, G. Adam and F. Berthiller, Critical evaluation of indirect methods for the determination of deoxynivalenol and its conjugated forms in cereals, *Anal. Bioanal. Chem.*, 2015, **407**, 6009–6020.
49. H. Michlmayr, E. Varga, A. Malachová, N. T. Nguyen, C. Lorenz, D. Haltrich, F. Berthiller and G. Adam, A Versatile Family 3 Glycoside Hydrolase from Bifidobacterium adolescentis Hydrolyzes β-Glucosides of the Fusarium Mycotoxins Deoxynivalenol, Nivalenol, and HT-2 Toxin in Cereal Matrices, *Appl. Environ. Microbiol.*, 2015, **81**, 4885–4893.
50. B. Poppenberger, F. Berthiller, H. Bachmann, D. Lucyshyn, C. Peterbauer, R. Mitterbauer, R. Schuhmacher, R. Krska, J. Glössl and G. Adam, Heterologous expression of Arabidopsis UDP-glucosyl-transferases in *Saccharomyces cerevisiae* for production of zearalenone-4-O-glucoside, *Appl. Environ. Microbiol.*, 2006, **72**, 4404–4410.
51. JECFA. Evaluation of certain contaminants in food: Seventy-second report of the Joint FAO/WHO Expert Committee on Food Additives. *WHO Technical Report Series*, 2011, 959.
52. F. Berthiller, R. Krska, K. J. Domig, W. Kneifel, N. Juge, R. Schuhmacher and G. Adam, Hydrolytic fate of deoxynivalenol-3-glucoside during digestion, *Toxicol. Lett.*, 2011, **206**, 264–267.
53. A. Dall'Erta, M. Cirlini, M. Dall'Asta, D. Del Rio, G. Galaverna and C. Dall'Asta, Masked mycotoxins are efficiently hydrolyzed by human

colonic microbiota releasing their aglycones, *Chem. Res. Toxicol.*, 2013, **26**, 305–312.

54. S. W. Gratz, G. Duncan and A. J. Richardson, The Human Fecal Microbiota Metabolizes Deoxynivalenol and Deoxynivalenol-3-Glucoside and May Be Responsible for Urinary Deepoxy-Deoxynivalenol, *Appl. Environ. Microbiol.*, 2013, **79**, 1821–1825.
55. V. Nagl, H. Schwartz, R. Krska, W. D. Moll, S. Knasmüller, M. Ritzmann, G. Adam and F. Berthiller, Metabolism of the masked mycotoxin deoxynivalenol-3-glucoside in rats, *Toxicol. Lett.*, 2012, **213**, 367–373.
56. V. Nagl, B. Woechtl, H. E. Schwartz-Zimmermann, I. Hennig-Pauka, W. D. Moll, G. Adam and F. Berthiller, Metabolism of the masked mycotoxin deoxynivalenol-3-glucoside in pigs, *Toxicol. Lett.*, 2014, **229**, 190–197.

Natural Occurrence of Masked Mycotoxins

COLIN CREWS* AND SUSAN JANE MacDONALD

Food and Environment Research Agency, Sand Hutton, York YO41 1LZ,
United Kingdom
*Email: colin.crews@fera.gsi.gov.uk

2.1 Introduction

Databanks containing information describing the mycotoxin content of
foods and feed are of considerable importance in assessing consumer exposure, for setting legislative maximum limits for particular products and
for devising mitigation measures. The main consequences of the existence of
masked mycotoxins in this regard are that the data collections are very likely
to underestimate the mycotoxin level.

The decisions to include masked mycotoxins in databases of occurrence
and exposure are based on evidence that some cereals contain toxins in
bound forms as a high percentage of the total mycotoxin present; this has
been most strongly demonstrated for the glucoside of deoxynivalenol (DON-
3-Glc). The Joint FAO/WHO Expert Committee on Food Additives (JECFA) has
published a decision that if DON-3-Glc released deoxynivalenol (DON) on
ingestion, the quantity of DON available as glucoside should be added to the
permitted maximum total daily intake.[1] This follows a decision to produce a
group provisional maximum tolerable daily intake (PMTDI) for DON of
$1\ \mu g\ kg^{-1}$ b.w. that includes its acetylated derivatives 3-acetyl DON and 15-
acetyl DON.[2] DON-3-Glc was not included in this group PMTDI, but it is
likely to be added at a future date; therefore, data obtained from surveys of

Issues in Toxicology No. 24
Masked Mycotoxins in Food: Formation, Occurrence and Toxicological Relevance
Edited by Chiara Dall'Asta and Franz Berthiller
© The Royal Society of Chemistry 2016
Published by the Royal Society of Chemistry, www.rsc.org

occurrence are of considerable and increasing importance. China has added DON-3-Glc and the acetylated forms of DON to its national food contaminant surveillance network for future monitoring.

2.2   Masked forms of DON

The most extensive published information on masked mycotoxin occurrence is that relating to DON-3-Glc. Reliable information is available for its occurrence in various countries in a reasonably wide range of foods and ingredients including cereals and processed products, especially bread and beer, and for its complex relationship with free DON. DON-3-Glc is the metabolite most often isolated from contaminated cereals including wheat, oats and barley, as well as their products such as processed cereals, malt and beer. Several surveys of masked mycotoxins have included the acetylated analogues of DON, 3- and 15-acetyl-DON. However, these compounds are readily amenable to extraction and detection by all methods used to determine free *Fusarium* toxins and are not considered to be masked in this context.

2.2.1   Occurrence

The results of several surveys of the DON-3-Glc content of cereal and cereal-based products have been published, and a summary of the findings of those carried out in different countries since 2007, which have included a relatively large number of samples, are presented in Table 2.1.

A high proportion (80%) of 116 cereal products sampled in the Czech Republic in 2010 contained DON-3-Glc, which was present over the range 5–72 μg kg^{-1} with mean values ranging from 15 to 35 μg kg^{-1}. DON-3-Glc was found in 16% of 17 products made from white flour, 32% of 36 products made from mixed flour that included wheat and rye and 20% of 34 cereal-based snacks.[7] The highest concentrations, exceeding 30 μg kg^{-1}, were detected in breakfast cereals and snacks. Only bakery products made from white flour contained DON-3-Glc in excess of 40 μg kg^{-1}, and the highest level (72 μg kg^{-1}) was measured in a single whole-grain slice product.

DON, DON-3-Glc, zearalenone and zearalenone glucosides were found in fibre- and bran-enriched bread, cornflakes, popcorn and oatmeal sold in Belgium. DON-3-Glc was present in equal or greater concentration than the unbound form.[4,15,16] DON-3-Glc was found in half of the fibre-enriched bread samples and breakfast cereals, 77% of oatmeal and over 90% of the popcorn samples.

DON and DON-3-Glc were measured in 22 cereal samples, a malt syrup and three malt extracts from nine different countries worldwide.[17] DON-3-Glc was detected in 21 cereal samples and DON in 22. The levels of DON-3-Glc ranged from <1 μg kg^{-1} to 367 μg kg^{-1}, with a median of 19 μg kg^{-1}. It was found in only one malt-based product (malt extract) at 6 μg kg^{-1}. The DON contents of wheat flours were generally between 6% and 15% in the glucoside form, whereas for Polish barley and oat flour the proportions of the glucoside form were 24% and 29%, respectively.

Table 2.1 Levels of DON-3-Glc ($\mu g\,kg^{-1}$) reported in major surveys.

Food	No.	Country	Year	Max	Mean	Ref.
Barley	65	Belgium	2012	–[a]	390	3
Bread (bran enriched)	52	Belgium	2010–2011	425	34	4
Bread (fibre enriched)	36	Belgium	2010–2011	103	21	4
Corn	204	China	2007–2008	–[a]	21	5
Cornflakes	61	Belgium	2010–2011	63	13	4
Dark beer	47	EU	2011	26	6.9	6
Flour	22	Czech Republic	2010	72	15	7
Flour (mixed) products	36	Czech Republic	2010	41	19	7
Flour (white) products	17	Czech Republic	2010	30	15	7
Maize	54	EU	2006	763	141	8
Maize	288	Belgium	2011	1,100	37	9
Maize	26	Burkina Faso	2010	n.d.[b]	0	10
Maize kernels	203	China	2008	499	66	11
Maize kernels	20	China	2009	93	23	11
Maize kernels	60	China	2010	495	73	11
Maize products	384	China	2009	844	76	11
Maize products	155	China	2010	128	26	11
Maize products	141	China	2012	39	11	11
Pale beer	217	EU	2011	81	6.7	6
Processed snacks	34	Czech Republic	2010	94	32	7
Various	30	Burkina Faso	2010	24	32	10
Wheat	93	Belgium	2012	–[a]	250	3
Wheat	192	China	2007–2008	–[a]	35	5
Wheat	192	Czech Republic	2011	21	31	7
Wheat	17	Czech Republic	2010	30	–[a]	7
Wheat	23	EU	2006	1070	393	8
Wheat	88	USA	2008	2	–[a]	12
Wheat	140	USA	2009	4	–[a]	12
Wheat	356	USA	2010	3	–[a]	12
Wheat	54	Serbia	2007	46	–[a]	13
Wheat	54	Serbia	2007	83	–[a]	13
Wheat beer	46	EU	2011	28	11.5	6
Wheat flour	30	China	2008	39	–[a]	14
Wheat kernel	162	China	2008	238	–[a]	14
Wheat products	291	China	2009	235	–[a]	14
Wheat products	125	China	2010	53	–[a]	14
Wheat products	89	China	2011	87	–[a]	14

[a]Not reported.
[b]Not detected.

Conjugated DON was measured in Australian cereal grains harvested over 3 years from 2009 to 2011.[18] It was found in up to almost 50% of samples, with lower contamination in 2010 than in the earlier and later harvests. No significant difference was observed between free DON and total DON contents. Analysis of 84 cereal-based food products, mostly from the UK and assembled into 25 composites, revealed DON-3-Glc in only two composites.[19]

Three significant surveys determined the contamination of cereals and cereal-based foods in China with DON-3-Glc. A total of 446 corn and wheat samples harvested in 2007 and 2008 collected from seven provinces of China were analysed.[5] Both corn and wheat samples contained DON-3-Glc, with median levels of 35 μg kg^{-1} (corn) and 21 μg kg^{-1} (wheat). This report was followed by a further survey of wheat and wheat-based products sampled from 24 provinces in China during 2008–2011.[14] Where high levels of free DON were detected, they were accompanied by lower but moderate concentrations of DON-3-Glc. The concentrations of DON-3-Glc ranged from 4 to 238 μg kg^{-1} (mean 52 μg kg^{-1}) in wheat kernels in 2008, from 3 to 39 μg kg^{-1} (mean 11 μg kg^{-1}) in wheat flour in 2008 and up to a maxima of 235, 53 and 87 μg kg^{-1} in wheat products in 2009, 2010 and 2011, respectively. The average relative ratio of DON-3-Glc to DON was 33% in wheat kernels and 10% in wheat flour in 2008, and $22 \pm 7\%$, $9 \pm 4\%$ and $14 \pm 7\%$ in wheat products in 2009, 2010 and 2011, respectively. DON-3-Glc was present in 47% of the wheat kernels and in 63% of the flour sampled in 2008. The incidence of contamination of wheat-based products varied with sampling year, being highest in 2009 (83%) and similar (58% and 47%, respectively) in 2010 and 2011. Samples of corn kernels and corn-based food products taken from 17 regions sampled in China over 4 years were reported by Wang *et al.*[20] DON-3-Glc was present in 33% of corn kernels in 2008, 60% in 2009, 65% in 2010 and 83% in 2011. The occurrence in corn-based products was similarly high (64–86%). DON-3-Glc was present in corn kernels over the ranges 3–499 μg kg^{-1} (mean 66 μg kg^{-1}) in 2008, 3–93 μg kg^{-1} (mean 23 μg kg^{-1}) in 2009, 3–495 μg kg^{-1} (mean 73 μg kg^{-1}) in 2010 and 3–10 μg kg^{-1} (mean 6 μg kg^{-1}) in 2011. It was present in corn-based products over the ranges 3–844 μg kg^{-1} (mean 76 μg kg^{-1}) in 2009, 3–128 μg kg^{-1} (mean 26 μg kg^{-1}) in 2010 and 3–39 μg kg^{-1} (mean 11 μg kg^{-1}) in corn-based products in 2011. The occurrence of DON-3-Glc had a consistently positive correlation with that of free DON for all of the samples taken in these Chinese surveys.

2.2.2 Ratios of DON-3-Glc to DON

There are many reports of the ratio of DON-3-Glc to DON, but as data have accumulated, it has become clear that the ratio varies considerably. Some data on a limited number of samples suggest that the relative proportion of DON-3-Glc to DON is fairly constant at about 20%,[18,21] rarely if ever exceeding 30% in cereals,[22] but the quantity of DON-3-Glc in beer has sometimes exceeded that of DON,[23] and it has in some samples exceeded the level of DON by almost three-fold.[24]

Typical ratios of DON-3-Glc to DON in Austrian wheat were about 10%, but some cultivars contained 30% DON-3-Glc.[21] DON-3-Glc in beer is normally present at a level compared to DON present, but occasionally levels are higher.[25,26] Li *et al.*[5] reported DON-3-Glc over a range of 7–56% of the DON content of some Chinese wheat cultivars. However, the ratio of DON-3-Glc to free DON was less than 1 in most of the samples analysed, although some higher values of DON-3-Glc than free DON were observed. The average DON-3-Glc/DON ratio in 697 samples of wheat and wheat-based products from

China over the 4 year sampling period of 2008–2011 was 21%, but the highest ratio was 1 to 1.570%,[14] much higher than those reported by Berthiller *et al.*[21] and Desmarchelier and Seefelder.[17]

The proportion of DON-3-Glc in 17 Czech Republic white flours was about 10%, with about 20% in mixed flour products and over 50% in cereal-based snacks.[7] Two samples of bread containing barley malt and malted sprouting barley also contained DON-3-Glc but no free DON.

2.2.3 Occurrence in Beer

DON-3-Glc has been found as a contaminant in a high proportion of a large number of beers analysed in several surveys. It was usually present at about 40% of the levels of DON. The results of a large survey of free and masked DON mycotoxins in 374 beers from 38 countries, but centred on those from Austria and Germany, showed that over 90% contained both DON-3-Glc and DON, although over 75% of the samples had less than 5 mg L^{-1} of both.[6] The molar ratio of DON-3-Glc to DON varied from 0.11 to 1.25 regardless of beer category, with an average of 0.6. Six categories of beer were analysed including pale, wheat, dark, non-alcoholic beers and minor groups. The average DON-3-Glc concentration was about 7 mg L^{-1}. The highest levels of both DON-3-Glc and DON (81 mg L^{-1} and 89 mg L^{-1}, respectively) were found in a pale Austrian beer (although only two Austrian beers exceeded 40 mg L^{-1} DON-3-Glc), and the lowest levels were in non-alcoholic beers and shandies, which contained less than 5 mg L^{-1}. The average level of DON-3-Glc across all of the categories of all samples was about 7 mg L^{-1} for DON-3-Glc, only slightly lower than that of DON (8.4 mg L^{-1}). Where present, the average DON-3-Glc content was 9.5 mg L^{-1}, again slightly lower than that of DON at 13.6 mg L^{-1}.

About 75% of samples (176) of a wide range of beers collected in the Czech Republic in 2007 contained detectable DON-3-Glc.[23] This was a greater incidence than that found for free DON. Few non-alcoholic beers contained DON-3-Glc or DON, perhaps because the processing mechanism designed to lower alcohol production also affected release of mycotoxins. When analysed by liquid chromatography–tandem mass spectrometry (LC-MS/MS), roasted malts did not appear to contain detectable DON-3-Glc, but analysis of the same samples using two enzyme-linked immunosorbent assay (ELISA) kits gave high but different responses, it being likely that the ELISAs had cross-reactions with other compounds.

Various isomeric forms of di- and tri-glucosides of DON have been reported in a beer sample,[27] but routine monitoring of these higher glucosides has not been put into practice.

2.2.4 Other DON Metabolites

Analysis by high-performance liquid chromatography (HPLC) with tandem mass spectrometry (MS/MS) has revealed the presence of ten biotransformation products of DON in the ears of corn inoculated with both native and isotopically labelled DON.[28] These comprised DON and DON-3-Glc and eight additional biotransformation products, one of which was putatively identified as a DON–glutathione conjugate, and others as DON adducts with cysteine and cysteinyl–glycine. No further monitoring of these or novel metabolites has been reported to date.

2.3 Masked Zearalenone

In the text and tables below, the uniform way of numbering ring positions and of abbreviating the names of zearalenone and its metabolites proposed by Metzler[29] has been adopted.

Plants infected with *Fusarium* that produce the toxin zearalenone can transform it into a range of glucose conjugates, principally isomers of zearalenol (α- and β-ZEL), α- and β-zearalenol-14-glucopyranoside (ZEN-14-Glc) and zearalenone-14-sulphate (ZEN-14-S).

In a survey of zearalenone and its derivatives in 84 cereal-based foods ZEN-14-S predominated. The samples comprised 13 types in composites including wheat flour, whole-meal wheat bread, maize meal, biscuits, wheat flakes, bran flakes, muesli, crackers, cereal snack bars and polenta. ZEN-14-S was present in 13 composites, with the highest quantity being about 6 µg kg^{-1} in bran flakes. The survey highlighted the wide distribution of ZEN-14-S in cereal products, it having been found in many different commodities, albeit in low concentrations.[19]

In a survey of extractable conjugated *Fusarium* mycotoxins in composites of cereal-based raw materials and finished products carried out by Vendl *et al.*,[19] none of 84 cereal-based product composites contained ZEN-14-Glc, α- or β-ZEL, α-zearalenone-glucopyranoside (αZEN-14-Glc) or β-zearalenone-glucopyranoside (βZEN-14-Glc). Masked forms of zearalenone were also determined in a survey of 174 cereal-based foods, 67 compound feeds and 19 raw feed materials purchased in Belgium during 2010 and 2011.[4] Levels of bound zearalenone were generally low, with the highest levels (up to 59 µg kg^{-1}) being in cornflakes during 2011, and the highest level in this food was 1 µg kg^{-1} in 2010. Data for masked forms of zearalenone and zearalenol measured in the Belgian survey are collated in Table 2.2.

Thirty samples of a variety of food and feed matrices including maize, wheat, oats, cornflakes and bread were analysed for ZEN, α- and β-ZEL, ZEN-14-Glc, α- and β-ZEL-14-Glc and ZEL-14-S.[15] The incidence of ZEN in food and feed matrices was 80%. αZEL and βZEL, respectively, occurred in 53% and 63% of the samples. ZEN-14-Glc was detected in nine samples, from trace levels up to 274 µg kg^{-1}. In one maize sample, the co-occurrence of ZEN-14-Glc (274 µg kg^{-1}), ZEN-14-S (51 µg kg^{-1}), βZEL-14-Glc (92 µg kg^{-1}) and the relatively low amount of ZEN (59 µg kg^{-1}) suggested that approximately 90% of the available ZEN was metabolised. In fibre- and bran-enriched bread, cornflakes, popcorn and oatmeal sold in Belgium,[16] ZEN-14-Glc, ZEN-14-S, αZEL-14-Glc and βZEL-14-Glc occurred in 29%, 8%, 10% and 19% of the fibre-enriched bread samples, respectively. In the bran-enriched

Table 2.2 Levels of conjugated zearalenol and zearalenone reported in Belgian food and feed.[4,16]

	Food	No.	Year	Max	Mean
ZEN-14-Glc	Bread (Fiber enriched)	52	2010–2011	154	15
ZEN-14-Glc	Bread (Bran enriched)	36	2010–2011	155	18
ZEN-14-Glc	Cornflakes	61	2010–2011	369	39
ZEN-14-Glc	Popcorn	12	2010–2011	n.d.[b]	n.d.[b]
ZEN-14-Glc	Oatmeal	13	2010–2011	91	12
ZEN-14-Glc	Poultry feed	14	2010–2011	282	28
ZEN-14-Glc	Piglet feed	8	2010–2011	164	28
ZEN-14-Glc	Sow feed	15	2010–2011	64	11
ZEN-14-Glc	Pig feed	13	2010–2011	36	8
ZEN-14-Glc	Horse feed	14	2010–2011	296	27
ZEN-14-S	Poultry feed	14	2010–2011	25	4
ZEN-14-S	Piglet feed	8	2010–2011	127	28
ZEN-14-S	Sow feed	15	2010–2011	38	11
ZEN-14-S	Pig feed	13	2010–2011	64	15
ZEN-14-S	Horse feed	14	2010–2011	47	4
Masked ZEN[a]	Maize	288	2011	9750	524
αZOL-14-Glc	Bread (bran enriched)	52	2010–2011	63	3
αZOL-14-Glc	Bread (fibre enriched)	36	2010–2011	12	0.3
αZOL-14-Glc	Cornflakes	61	2010–2011	192	11
αZOL-14-Glc	Popcorn	12	2010–2011	n.d.[b]	n.d.[b]
αZOL-14-Glc	Oatmeal	13	2010–2011	10	1
αZOL-14-Glc	Poultry feed	14	2010–2011	75	11
αZOL-14-Glc	Piglet feed	8	2010–2011	75	11
αZOL-14-Glc	Sow feed	15	2010–2011	96	9
αZOL-14-Glc	Pig feed	13	2010–2011	418	5
αZOL-14-Glc	Horse feed	14	2010–2011	n.d.[b]	0
βZOL-14-Glc	Bread (Fiber enriched)	52	2010–2011	153	7
βZOL-14-Glc	Bread (Bran enriched)	36	2010–2011	153	6
βZOL-14-Glc	Cornflakes	61	2010–2011	206	11
βZOL-14-Glc	Popcorn	12	2010–2011	10	1
βZOL-14-Glc	Oatmeal	13	2010–2011	10	4
βZOL-14-Glc	Poultry feed	14	2010–2011	72	15
βZOL-14-Glc	Piglet feed	8	2010–2011	35	14
βZOL-14-Glc	Sow feed	15	2010–2011	53	10
βZOL-14-Glc	Pig feed	13	2010–2011	44	37
βZOL-14-Glc	Horse feed	14	2010–2011	21	3

[a]Sum of zearalenone-14-glucoside, zearalenone-14-sulphate, α-zearalenol-14-glucoside and β-zearalenol-14-glucoside.
[b]Not detected.

bread samples, ZEN-14-S occurred in 6% of the samples, and the glucosylated forms occurred in 6% (ZEN-14-Glc), 3% (αZEL-14-Glc) and 6% (βZEL-14-Glc). Concerning the breakfast samples, αZEL-14-Glc, βZEL-14-Glc and ZEN-14-S occurred in the same incidence (26%, 29% and 27%), while ZEN-14-Glc was observed in 40% of the samples. None of the popcorn samples contained αZEL-14-Glc, nor ZEN-14-Glc. All other masked forms also occurred only in very small amounts (8%). Only two oatmeal samples were contaminated with ZEN-4-S and βZEL-14-Glc, and ZEN-14-Glc was present in 38% of the samples.

Fewer studies have been devoted to crop plants. Schneweis *et al.*[30] demonstrated the presence of zearalenone metabolites in a survey of 10 wheat grain samples, where the relative proportion of ZEN-14-Glc to ZEN was found to be on average about 27%.

2.4 Masked Fumonisins

Fusarium verticillioides and *Fusarium proliferatum*, the two fungi mainly responsible for pink ear rot on maize, produce fumonisins. Fumonisin B_1 is the most prevalent, which commonly co-occurs with fumonisin B_2 and B_3, although other types have been reported. Fumonisin B_1 is a long-chain hydroxylated alkylamine esterified with two propane-1,2,3-carboxylic acid (tricarballylic acid [TCA]) side chains, and the other fumonisins are structurally similar.

The occurrence of masked or 'modified' fumonisin B_1 has been known since the early 1990s, a relatively short time after the discovery and structural elucidation of the fumonisins. The occurrence of hydrolysed fumonisin B_1 (HFB_1), also known as aminopentol (AP_1), was reported in alkaline-treated maize, typically following the nixtimalisation process using calcium hydroxide that is used in the production of traditional tortillas in South America.[31] The alkaline treatment removes the TCA side chains, leaving the hydrolysed product. However, HFB_1 was also reported in other products not treated with alkali, including canned yellow corn,[31] while naturally occurring partially hydrolysed fumonisin B1 ($PHFB_1$) was reported by Xie *et al.*[32] to co-occur with fumonisin B_1 in corn and corn screenings. This meant from an early stage that two potential mechanisms for the formation of masked fumonisins were known, one *via* food production processes and the other *via* a naturally occurring phenomenon, although whether this was *via* plant or fungal metabolism was not known.

Some initial work to identify thermal conversion products of FB_1 was carried out by Shier *et al.*[33] The major product found at temperatures commonly used during roasting or baking was an anhydride (loss of one water molecule). This was confirmed by fast atom bombardment mass spectrometry. It was expected that this anhydride would react with thiol groups in proteins to produce covalently bound derivatives. Deuterium-labelled HFB_1 (^3H-HFB_1) was used in further studies to trace the fate of HFB_1 in simulated frying and roasting conditions. This confirmed that frying in oil produced less polar derivatives, while both processes resulted in approximately 46% of the radiolabel being detected in the protein-containing fraction for the roasted conditions and about 26% for the fried material. In both cases, it was shown that the linkage was hydrolysable, although the products were not identified.

It was also demonstrated by model experiments that fumonisin B_1 can bind to polysaccharides and proteins *via* their TCA side chains.[34] However, this model explains that fumonisin binding to protein or carbohydrate can only occur through heating and so cannot explain the similar binding observed in raw maize. In this case, the binding that occurs is more likely to be physical entrapment either by starch or protein supramolecular structures. This may be due to enzymatic activity or some other unknown phenomenon.[35] Using a digestion protocol, evidence of this problem was clearly demonstrated by the analysis of a reference material with a declared fumonisin content of $3036 \pm 746\ \mu g\,kg^{-1}$, whereas after digestion, a level as high as $8010 \pm 426\ \mu g\,kg^{-1}$ was detected. This paper also suggested clarification of the terminology used, as until that time the terms 'hidden' and 'bound' fumonisins had been used interchangeably. It was proposed to use 'bound' only for those compounds with a covalent linkage and 'hidden' for non-covalently bound derivatives formed *via* an associative interaction with the matrix.[36]

Three fatty acid esters of fumonisin B_1 have been identified from cultures of *F. verticillioides*-infected solid rice using HPLC with ion-trap mass spectrometry (ITMS) and HPLC with time-of-flight mass spectrometry. The fatty acids were linoleic acid (LA), palmitic acid (PA) and oleic acid (OA). Three pairs of isomers were identified that the authors denoted as esterified FB_1 (EFB_1) toxins, with the suggested names EFB_1 PA, iso-EFB_1 PA, EFB_1 LA, iso-EFB_1 LA, EFB_1 OA and iso-EFB_1 OA.[36] More recently, the presence of EFB_1 LA and EFB_1 OA has been reported in raw maize for the first time. In the same study, it was shown that esterification occurred in cultures of *F. verticillioides* grown on malt extract and corn meal-based growth media. Esters were only produced in the corn-based media, suggesting that esterification may only occur in complex matrices such as corn.[37] Studies on different maize hybrids found a correlation with higher fumonisin contamination in hybrids showing a higher LA content and a higher masking action in hybrids with a higher OA to LA ratio. Fatty acids may be implicated in the masking phenomenon.

Despite the research work and model studies carried out to understand the mechanisms of interaction between fumonisins and matrix, there have been very few published large studies on the occurrence of hidden or bound fumonisins in foods.

The first report of 'hidden' fumonisins in a commercial food product was in 2003 when Kim *et al.* reported the occurrence of HFB_1 in cornflakes.[38] Samples of cornflakes that had been analysed by traditional C18 and immunoaffinity column clean-up steps were further extracted using a solution of sodium dodecyl sulphate (SDS), followed by a hydrolysis step. HFB_1 was measured using HPLC with fluorescence detection. In all cases, even when no fumonisins had been found, HFB_1 was detected. The authors stated that this showed the presence of protein-bound fumonisin B_1, with on average 2.6 times more HFB_1 present than FB_1. Further work carried out adapted the original method, as the whole sample was subjected to hydrolysis with KOH as well as extraction with SDS to determine protein-bound fumonisin. Of 15 breakfast cereals and snacks analysed, 14 contained FB_1, while all contained some bound fumonisin. The incidence of any fumonisins was lower in the alkali-treated foods, with seven samples containing FB_1 and six samples containing HFB_1. Compared with FB_1 determined by traditional analysis, about 1.3 and 0.9 times more FB_1 was detected in bound forms in corn breakfast cereals and alkali-processed corn foods, respectively.[39]

A survey of gluten-free products on the Italian market found some bound fumonisins in all samples tested at similar or higher levels than the free forms. Some samples were over the maximum EU legal limit for foods for human consumption for total fumonisins.[35] More recently, an *in vitro* digestion assay approach found that hidden fumonisin was equal to or greater than the amount of free fumonisin measured. This has implications for consumers who could be routinely exposed to higher than anticipated levels of fumonisins.[40] The European Food Safety Authority (EFSA) recently evaluated modified mycotoxins and published a scientific opinion that included fumonisins. The panel assessed the exposure of modified forms in addition to parent forms. In the case of fumonisins, 60% was added to account for additional hidden and bound fumonisins in the absence of published occurrence or exposure data on modified forms.[41] It was concluded that, for fumonisins and modified fumonisins, the exposure of toddlers and other children exceeded the PMTDI at both the Lower Bound and the Upper Bound estimates, which could be of concern.[41]

2.5   Ochratoxin A

The metabolism of ochratoxin A (OTA) has been studied using cell suspension cultures of several plants, including incubation with the radiolabelled toxin.[42] The experiments showed that OTA conversion was almost quantitative, with the main metabolites isolated being optical isomers of 4-hydroxy-OTA and their glucosides. Hydroxy-OTA has not been reported in surveys; however, findings of the glucosides are beginning to appear. OTA is lost from infected coffee beans during roasting. About half of the toxin is degraded to identifiable products, but it has been demonstrated experimentally that OTA can be bound at roasting temperature by esterification to polysaccharides such as those found in coffee.[43] Ochratoxin B has been described as one of a few products of coffee roasting that can give a false-positive reaction on ELISA determination.[44]

2.6   T2 and HT2 Toxins

The existence of mono-glucosylated derivatives of the *Fusarium* toxins T2 and HT2 has been reported for corn,[45] a finding that was followed up by the discovery of di- and tri-glucosylated forms.[46] The presence of the more highly glucosylated forms was itself hidden by a lack of knowledge of the type of sodium and ammonium adduct formation on LC-MS analysis. Di-glucosylated HT2 was reported in two of 20 samples of barley, wheat and oats naturally infected with *Fusarium*,[47] and in wheat and oats.[48] The latter authors detected the isomeric HT2-3-glucoside and HT2-4-glucoside in addition to T2 and HT2 monoglucosides. Semi-quantitative analyses also indicated that the occurrence of T2-Glc and HT2-Glc was more likely in wheat than in oats. The range and degree of occurrence of T2 and HT2 glucosides in cereals and in cereal products has not yet been surveyed on a sufficient scale to draw conclusions on the level of human or animal exposure. T2 and HT2 glucosides were detected in bread made from grain contaminated with T2 and HT2. Semi-quantitative determination suggested that levels of HT2 glucoside decreased on baking bread, whereas T2 glucoside increased.[49]

2.7 Fusarenon-X

A new *Fusarium* mycotoxin glucoside, fusarenon-X-glucoside (FUSX-Glc), has recently been reported for the first time in wheat grain that was artificially infected with *Fusarium* fungi. Another mycotoxin glucoside, nivalenol-glucoside, was also found in the same grain sample. The authors estimated that more than 15% of fusarenon-X and nivalenol were converted into their respective glucosides.[45] Wheat and oats inoculated with *Fusarium sporotrichiodes* have also been found to contain 3-*O*-glucosides of T2 toxin and HT2 toxin, strongly suggesting the natural occurrence of these compounds as well.[50]

FUSX-Glc and nivalenol glucoside (NIV-Glc) have been detected in wheat grain that was artificially infected with *Fusarium* fungi. Over 15% of the fusarenon-X and nivalenol were present as their glucosides.[45] It is probable that the 3-OH glucoside as fusarenon has a structure similar to that of DON, which is conjugated mainly as DON-3-Glc.

Glucosides identified as being derived from the type A trichothecenes neosolaniol-glucoside (NEO-Glc) and diacetoxyscirpenol-glucoside (DAS-Glc) have been found in a commercially available corn powder reference material.[46]

Transformation of the neurotoxin fusaric acid to its *N*-methylamide derivative by many plant species was documented over 50 years ago (reviewed in Karlovsky[51]).

2.8 Other Masked Mycotoxins

One study reported the possible occurrence of bound patulin in apple juice.[52] In particular, a decrease in patulin recovery was observed during storage when cloudy apple juice was spiked with this mycotoxin. The decrease was significantly more pronounced for lower spiking levels. The authors hypothesised an interaction between the solid part of the juice and patulin. This was also in agreement with the previously reported observation that patulin contamination of cloudy apple juice can be reduced upon clarification and that the solid residue becomes enriched with patulin.[53] Since patulin is able to undergo an electrophilic attack on molecules containing a nucleophilic group, in particular with proteins or small peptides containing cysteine, lysine or histidine residues,[54] a binding between this compound and the solid part of cloudy apple juice may be supposed.

Some macrocyclic trichothecenes, including verrucarin-A and roridines A, D, and E, have been found as glucosides in the poisonous plant *Baccharis coridifolia*,[55,56] where they might act as feeding deterrents to herbivores.

Some studies have reported the possibility of acyl conjugation of mycotoxins in plants. The synthesis of these types of conjugates could be catalysed by acyltransferases. Acyl conjugates such as palmitoyl trichothecolone, palmitoyl scirpentriol and palmitoyl T-2 have been described after a natural and artificial infection of banana with *F. verticillioides* (syn. *F. moniliforme*),[57] but these results have been refuted by others.[58] Another example is the cinnamic acid ester of trichothecolone in anise seeds infected with *Trichothecium roseum*.[59]

Particular means of food processing might affect the incidence and level of masked mycotoxin formation, but to establish this effect, studies must be done into the contamination of the raw materials and of the products made from these materials, which is not usually carried out in surveys.[7] The masked mycotoxin content of processed foods is probably dependent on the level of conversion of free to bound forms in the living crop and the type of processing as it affects enzymatic processes. Several studies have been conducted into the effects of food processing operations on DON-3-Glc, but in many cases, findings differ because the conditions applied under these processes vary greatly according to manufacture and starting material and any trends are difficult to identify.

Effects of food processing on the levels of (masked) mycotoxins are described in detail in Chapter 5 of this book.

2.9   Effect of Climate on Occurrence

Recent studies in the USA have suggested that environmental conditions, weather and growing region have significant effects on both DON-3-Glc and DON in wheat.[60] The environmental conditions determine the degree of kernel damage and subsequent *Fusarium* infection. Within a year, levels of DON were correlated significantly with the kernel quality and the level of DON-3-Glc.

The distribution of masked mycotoxins across different parts of cereal plants has not yet been studied in much detail. For example, Schollenberger *et al.*[61] found that mycotoxin contamination of maize plants varied with the toxin type, with significant differences existing between the parts of the plant. This effect is likely to be equally true for the masked forms of the mycotoxins.

2.10   Conclusion

Data describing the occurrence of masked forms of the common mycotoxins are becoming more widely available and the scope of surveys in terms of geographical location, variety of samples and numbers of samples is increasing. Our knowledge of human exposure is, however, limited by the variety of analytical methods used and the lack of validation of these methods by large-scale collaborative trials. The situation will undoubtedly improve considerably in the future as increased interest stimulates method testing, and potential legislation, industry guidance and trading standards lead to an expansion of analytical activity. Hopefully, this will be augmented by the availability at some date of certified reference materials, reference standards and suitable internal standards.

References

1. Joint FAO/WHO Expert Committee on Food Additives (JECFA). Safety evaluation of certain contaminants in food. Prepared by the seventy-second meeting of the Joint FAO/WHO Expert Committee on Food Additives, 2011, WHO Food Additives Series 63. World Health Organization, Geneva, Switzerland. Available at: http://whqlibdoc.who.int/publications/2011/9789241660631_eng.pdf.

2. Joint FAO/WHO Expert Committee on Food Additives. Seventy-second meeting Rome, 16–25 February 2010, Summary and conclusions. Issued 16th March 2010.

3. A. Vanheule, K. Audenaert, M. De Boevre, S. Landschoot, B. Bekaert, F. Munaut, M. Eeckhout, M. Höfte, S. De Saeger and G. Haesaert, The compositional mosaic of *Fusarium* species and their mycotoxins in unprocessed cereals, food and feed products in Belgium, *Int. J. Food Microbiol.*, 2014, **181**, 28–36.

4. M. de Boevre, J. Diana di Mavungu, S. Landschoot, K. Audenaert, M. Eeckhout, P. Maene, G. Haesaert and S. de Saeger, Natural occurrence of mycotoxins and their masked forms in food and feed products, *World Mycotoxin J.*, 2012, **5**, 207–219.

5. F.-Q. Li, C.-C. Yu, B. Shao, W. Wang and H.-X. Yu, Natural occurrence of masked deoxynivalenol and multi-mycotoxins in cereals from China harvested in 2007 and 2008, *Zhonghua Yu Fang Yi Xue Za Zhi [Chinese Journal of Preventive Medicine]*, 2011, **45**, 57–63.

6. E. Varga, A. Malachova, H. Schwartz, R. Krska and F. Berthiller, Survey of deoxynivalenol and its conjugates deoxynivalenol-3-glucoside and 3-acetyl-deoxynivalenol in 374 beer samples, *Food Addit. Contam., Part A*, 2013, **30**, 137–146.

7. Z. Malachova, Z. Dzuman, M. Veprikova, M. Vaclavikova, M. Zachariasova and J. Hajslova, Deoxynivalenol, Deoxynivalenol-3-glucoside, and Enniatins: The Major Mycotoxins Found in Cereal-Based Products on the Czech Market, *J. Agric. Food Chem.*, 2011, **59**, 12990–12997.

8. F. Berthiller, C. Dall'Asta, R. Corradini, R. Marchelli, M. Sulyok, R. Krska, G. Adam and R. Schuhmacher, Occurrence of deoxynivalenol and its 3-β-D-glucoside in wheat and maize, *Food Addit. Contam., Part A*, 2009, **26**, 507–511.

9. M. De Boevre, S. Landschoot, K. Audenaert, P. Maene, J. Diana Di Mavungu, M. Eeckhout, G. Haesaert and S. De Saeger, Occurrence and within field variability of *Fusarium* mycotoxins and their masked forms in maize crops in Belgium, *World Mycotoxin J.*, 2014, 7, 91–102.

10. B. Warth, B. Parich, J. Atehnkeng, R. Bandyopadhyay, R. Schuhmacher, M. Sulyok and R. Krska, Quantitation of mycotoxins in food and feed from Burkina Faso and Mozambique using a modern LC-MS/MS multitoxin method, *J. Agric. Food Chem.*, 2012, **60**, 9352–9363.

11. W. Wei, M. Jiao-Jie, Y. Chuan-Chuan, L. Xiao-Hui, J. Hong-Ru, S. Bing and L. Feng-Qin, Simultaneous determination of masked deoxynivalenol and some important Type B Trichothecenes in Chinese corn kernels and corn-based products by Ultra-Performance Liquid Chromatography-Tandem Mass Spectrometry, *J. Agric. Food Chem.*, 2012, **60**, 1638–11646.

12. M. Ovando-Martínez, B. Ozsisli, J. Anderson, K. Whitney, J.-B. Ohm and S. Simsek, Analysis of deoxynivalenol and deoxynivalenol-3-glucoside in hard red spring wheat inoculated with *Fusarium graminearum*, *Toxins*, 2013, **5**, 2522–2532.

13. B. Skrbic, A. Malachova, J. Zivancev, Z. Veprikova and J. Hajslová, Fusarium mycotoxins in wheat samples harvested in Serbia: A preliminary survey, *Food Control*, 2011, **22**, 1261–1267.

14. F. Q. Li, W. Wang, J. J. Ma, C. C. Yu, X. H. Lin and W. X. Yan, , Natural occurrence of masked deoxynivalenol in Chinese wheat and wheat-based products during 2008–2011, *World Mycotoxin J.*, 2012, **5**, 221–230.

15. M. De Boevre, J. Diana di Mavungu, P. Maene, K. Audenaert, D. Deforce, G. Haesaert, M. Eeckhout, A. Callebaut, F. Berthiller, C. Van Peteghem and S. De Saeger, Development and validation of an LC-MS/MS method for the simultaneous determination of deoxynivalenol, zearalenone, T-2-toxin and some masked metabolites in different cereals and cereal-derived food, *Food Addit. Contam., Part A*, 2012, **29**, 819–835.

16. M. de Boevre, L. Jacxsens, C. Lachat, M. Eeckhout, J. D. di Mavungu, K. Audenaert, P. Maene, G. Haesaert, P. Kolsteren, B. de Meulenaer and S. de Saeger, Human exposure to mycotoxins and their masked forms through cereal-based foods in Belgium, *Toxicol. Lett.*, 2013, **218**, 281–292.

17. A. Desmarchelier and W. Seefelder, Survey of deoxynivalenol and deoxynivalenol-3-glucoside in cereal-based products by liquid chromatography electrospray ionization tandem mass spectrometry, *World Mycotoxin J.*, 2011, **4**, 29–35.

18. S. T. Tran and T. K. Smith, A survey of free and conjugated deoxynivalenol in the 2008 corn crop in Ontario, Canada, *Anim. Feed Sci. Technol.*, 2011, **163**, 84–92.

19. O. Vendl, C. Crews, S. MacDonald, R. Krska and F. Berthiller, Simultaneous determination of deoxynivalenol, zearalenone, and their major masked metabolites in cereal-based food by LC-MS-MS, *Food Addit. Contam. Part A*, 2010, **27**, 1148–1152.

20. W. Wang, J.-J. Ma, C.-C. Yu, X.-H. Lin, H.-R. Jiang, B. Shao and F.-Q. Li, Simultaneous Determination of Masked Deoxynivalenol and Some

Important Type B Trichothecenes in Chinese Corn Kernels and Corn-Based Products by Ultra-Performance Liquid Chromatography-Tandem Mass Spectrometry, *J. Agric. Food Chem.*, 2012, **60**, 11638–11646.

21. F. Berthiller, C. Dall'Asta, R. Schuhmacher, M. Lemmens, G. Adam and R. Krska, Masked mycotoxins: determination of a deoxynivalenol glucoside in artificially and naturally contaminated wheat by liquid chromatography-tandem mass spectrometry, *J. Agric. Food Chem.*, 2005, **53**, 3421–3425.

22. G. Galaverna, C. Dall'Asta, M. A. Mangia, A. Dossena and R. Marchelli, Masked Mycotoxins: an Emerging Issue for Food Safety, *Czech J. Food Sci.*, 2009, **27**, S89–S92.

23. M. Kostelanska, J. Hajslova, M. Zachariasova, A. Malachova, K. Kalachova, J. Poustka, J. Fiala, P. M. Scott, F. Berthiller and R. Krska, Occurrence of deoxynivalenol and its major conjugate, deoxynivalenol-3-glucoside, in beer and some brewing intermediates, *J. Agric. Food Chem.*, 2009, **57**, 3187–3194.

24. J. J. Sasanya, C. Hall and C. Wolf-Hall, Analysis of deoxynivalenol, masked deoxynivalenol, and *Fusarium graminearum* pigment in wheat samples, using liquid chromatography-UV-mass spectrometry, *J. Food Prot.*, 2008, **71**, 1205–1213.

25. F. Berthiller, R. Schuhmacher, G. Adam and R. Krska, Formation, determination and significance of masked and other conjugated mycotoxins, *Anal. Bioanal. Chem.*, 2009, **395**, 1243–1252.

26. F. Berthiller, C. Crews, C. Dall'Asta, S. De Saeger, G. Haesaert, P. Karlovsky, I. P. Oswald, W. Seefelder, G. Speijers and J. Stroka, Masked Mycotoxins: A Review, *Mol. Nutr. Food Res.*, 2013, **57**, 165–186.

27. M. Zachariasova, T. Cajka, M. Godula, M. Kostelanska, A. Malachova, and J. Hajslova, *Book of Abstracts of the 4th International Symposium on Recent Advances in Food Analysis (RAFA)*, Institute of Chemical Technology Prague, 2009, p. 391.

28. B. Kluger, C. Bueschl, M. Lemmens, F. Berthiller, G. Häubl, G. Jaunecker, G. Adam, R. Krska and R. Schuhmacher, Stable isotopic labelling-assisted untargeted metabolic profiling reveals novel conjugates of the mycotoxin deoxynivalenol in wheat, *Anal. Bioanal.Chem.*, 2013, **405**, 5031–5036.

29. M. Metzler, Proposal for a uniform designation of zearalenone and its metabolites, *Mycotoxin Res.*, 2011, **27**, 1–3.

30. I. Schneweis, K. Meyer, G. Engelhardt and J. Bauer, Occurrence of zearalenone-4-beta-D-glucopyranoside in wheat, *J. Agric. Food Chem.*, 2002, **50**, 1736–1738.

31. E. C. Hopmans and P. A. Murphy, Detection of fumonisins B_1, B_2, and B_3 and hydrolysed fumonisin B_1 in corn-containing foods, *J. Agric. Food Chem.*, 1993, **41**, 1655–1658.

32. W. Xie, C. J. Mirocha and J. Chen, Detection of two naturally occurring structural isomers of partially hydrolysed fumonisin B_1 in corn by

on-line capillary liquid chromatography-fast atom bombardment mass spectroscopy, *J. Agric. Food Chem.*, 1997, **45**, 1251–1255.

33. W. T. Shier, H. K. Abbas and F. A. Badria, Structure-activity relationships of the corn fungal toxin fumonisin B_1: implications for food safety, *J. Nat. Toxins*, 1997, **6**, 225–242.

34. W. Seefelder, A. Knecht and H. U. Humpf, Bound fumonisin B_1: analysis of fumonisin-B_1 glyco and amino acid conjugates by liquid chromatography-electrospray ionization-tandem mass spectrometry, *J. Agric. Food Chem.*, 2003, **51**, 5567–5573.

35. C. Dall'Asta, M. Mangia, F. Berthiller, A. Molinelli, M. Sulyok, R. Schuhmacher, R. Krska, G. Galaverna, A. Dossena and R. Marchelli, Difficulties in fumonisin determination: the issue of hidden fumonisins, *Anal. Bioanal. Chem.*, 2009, **395**, 1335–1345.

36. T. Bartók, L. Tölgyesi, Á. Mesterházy, M. Bartók and Á. Szécsi, Identification of the first fumonisin mycotoxins with three acyl groups by ESI-ITMS and ESI-TOFMS following RP-HPLC separation: palmitoyl, linoleoyl and oleoyl EFB1 fumonisin isomers from a solid culture of *Fusarium verticillioides*, *Food Addit. Contam., Part A*, 2010, **27**, 1714–1723.

37. C. Falavigna, I. Lazzaro, G. Galaverna, P. Battilani and C. Dall'Asta, Fatty acid esters of fumonisins: first evidence of their presence in maize, *Food Addit. Contam., Part A*, 2013, **30**, 1606–1613.

38. E. K. Kim, P. M. Scott and B. P. Y. Lau, Hidden fumonisin in corn flakes, *Food Addit. Contam.*, 2003, **20**, 161–169.

39. J. W. Park, P. M. Scott, B. P. Y. Lau and D. A. Lewis, Analysis of heat processed corn foods for fumonisins and bound fumonisins, *Food Addit. Contam.*, 2004, **21**, 1168–1178.

40. C. Dall'Asta, G. Falavigna, G. Galavern, A. Dossen and R. Marchelli, *In vitro* digestion assay for determination of hidden fumonisins in maize, *J. Agric. Food Chem.*, 2010, **58**, 12042–12047.

41. EFSA 2014, CONTAM Panel, Scientific Opinion on the risks for human and animal health related to the presence of modified forms of certain mycotoxins in food and feed, *EFSA J.*, 2014, **12**(3916), 107.

42. M. Ruhland, G. Engelhardt, W. Schaefer and P. R. Wallnöfer, Transformation of the mycotoxin ochratoxin A in plants: 1. Isolation and identification of metabolites formed in cell suspension cultures of wheat and maize, *Nat. Toxicol.*, 1996, **4**, 254–260.

43. A. Bittner, B. Cramer and H.-U. Humpf, Matrix binding of ochratoxin A during roasting, *J. Agric. Food Chem.*, 2013, **61**, 12737–12743.

44. M. Tozlovanu and A. Pfohl-Leszkowicz, Ochratoxin A in roasted coffee from French supermarkets and transfer in coffee beverages: comparison of analysis methods, *Toxins*, 2010, **2**, 1928–1942.

45. H. Nakagawa, K. Ohmichi, S. Sakamoto, Y. Sago, M. Kushiro, H. Nagashima, M. Yoshida and T. Nakajima, Detection of a new *Fusarium* masked mycotoxin in wheat grain by high-resolution LC-Orbitrap-MS, *Food Addit. Contam., Part A*, 2011, **28**, 1447–1456.

46. H. Nakagawa, S. Sakamoto, Y. Sago, M. Kushiro and H. Nagashima, Detection of masked mycotoxins derived from type A trichothecenes in corn by high-resolution LC-Orbitrap mass spectrometer, *Food Addit. Contam. Part A*, 2013, **30**, 1407–1414.
47. Z. Veprikova, M. Vaclavikova, O. Lacina, Z. Dzuman, M. Zachariasova and J. Hajslova, Occurrence of mono- and di-glycosylated conjugates of T-2 and HT-2 toxins in naturally contaminated cereals, *World Mycotoxin J.*, 2012, **5**, 231–240.
48. V. M. T. Lattanzio, M. Solfrizzo and A. Visconti, Enzymatic hydrolysis of T-2 toxin for the quantitative determination of total T-2 and HT-2 toxins in cereals, *Anal. Bioanal. Chem.*, 2009, **395**, 1325–1334.
49. E. De Angelis, L. Monaci, M. Pascale and A. Visconti, Fate of deoxynivalenol, T-2 and HT-2 toxins and their glucoside conjugates from flour to bread: an investigation by high-performance liquid chromatography high-resolution mass spectrometry, *Food Addit. Contam., Part A*, 2013, **30**, 345–355.
50. M. Busman, S. M. Poling and C. M. Maragos, Observation of T-2 Toxin and HT-2 Toxin glucosides from *Fusarium sporotrichioides* by liquid chromatography coupled to tandem mass spectrometry (LC-MS/MS), *Toxins*, 2011, **3**, 1554–1568.
51. P. Karlovsky, Biological detoxification of fungal toxins and its use in plant breeding, feed and food production, *Nat. Toxins*, 1999, **7**, 1–23.
52. K. Baert, B. De Meulenaer, C. Kasase, A. Huyghebaert, W. Ooghe and F. Devlieghere, Free and bound patulin in cloudy apple juice, *Food Chem.*, 2007, **100**, 1278–1282.
53. J. Bissessur, K. Permaul and B. Odhav, Reduction of patulin during apple juice clarification, *J. Food Prot.*, 2001, **64**, 1216–1219.
54. R. Fliege and M. Metzler, Electrophilic properties of patulin. Adduct structures and reaction pathways with 4-bromothiophenol and other model nucleophiles, *Chem. Res. Toxicol.*, 2000, **13**, 373–381.
55. B. B. Jarvis, N. Mokhtarirejali, E. P. Schenkel, C. S. Barros and N. I. Matzenbacher, Trichothecene mycotoxins from Brazilian Baccharis species, *Phytochemistry*, 1991, **30**, 789–797.
56. L. Rosso, M. S. Maier and M. D. Bertoni, Trichothecenes Production by the Hypocrealean Epibiont of Baccharis coridifolia, *Plant Biol.*, 2000, **2**, 684–686.
57. D. K. Chakrabarti and S. Ghosal, Occurrence of free and conjugated 12,13-epoxytrichothecenes and zearalenone in banana fruits infected with *Fusarium moniliforme*, *Appl. Environ. Microbiol.*, 1986, **51**, 217–219.
58. C. J. Mirocha, H. K. Abbas and R. F. Vesonder, Absence of trichothecenes in toxigenic isolates of *Fusarium moniliforme*, *Appl. Environ. Microbiol.*, 1990, **56**, 520–525.
59. S. Ghosal, D. K. Chakrabarti, A. K. Srivastava and R. S. Srivastava, Toxic 12,13-epoxytrichothecenes from anise fruits infected with *Trichothecium roseum*, *J. Agric. Food Chem.*, 1982, **30**, 106–109.

60. S. Simsek, K. Burgess, K. L. Whitney, Y. Gu and S. Y. Qian, Analysis of deoxynivalenol and deoxynivalenol-3-glucoside in wheat, *Food Control*, 2012, **26**, 287–292.
61. M. Schollenberger, H.-M. Müller, K. Ernst, S. Sondermann, M. Liebscher, C. Schlecker, G. Wischer, G. W. Drochner, K. Hartung and H.-P. Piepho, Occurrence and distribution of 13 trichothecene toxins in naturally contaminated maize plants in Germany, *Toxins*, 2012, **4**, 778–787.

Immunologically-based Methods for Detecting Masked Mycotoxins

CHRIS M. MARAGOS

USDA-ARS-NCAUR, 1815 N. University Street, Peoria, IL 61604, USA
Email: chris.maragos@ars.usda.gov

3.1 Introduction

3.1.1 Terminology

When it comes to immunoassays for masked mycotoxins, understanding the terminology (see also the introductory chapter of this book) is important for understanding what the assays are actually detecting. Rather than focusing on whether the product is a targeted or an untargeted analyte, or upon how it came to be present in the sample, this review chapter will focus on the forms that these materials take and their impact on mycotoxin immunoassays. Generally, the terms "covalent" or "non-covalent" will be used to indicate the relationship between the parent toxin and its derivative. As noted by Berthiller *et al.*,[1] those mycotoxins that are non-extractable completely elude conventional analysis. In this review, "extractable" or "non-extractable" will be used to indicate the relationship between the derivative and the matrix (commodity, food, plant material, fermentation culture, *etc.*). Non-extractable complexes, for example toxins physically entrapped within a matrix ("hidden" mycotoxins), might be rendered extractable through

Issues in Toxicology No. 24
Masked Mycotoxins in Food: Formation, Occurrence and Toxicological Relevance
Edited by Chiara Dall'Asta and Franz Berthiller

Published by the Royal Society of Chemistry, www.rsc.org

enzymatic or chemical techniques that disrupt the matrix or release covalently attached derivatives.[3]

3.1.2 Why use Immunoassays to Detect Masked Forms?

The number of mycotoxins and their potential precursors, metabolites and conjugates leaves analysts with a bewildering array of potential analytes. Which are the most significant? Significance (relevance) is revealed through toxicity testing and, at least initially, approximations of potential exposure. There are many ways to deal with such diversity. One approach is to develop assays to detect each of the individual toxins. Advanced mass spectrometry (MS) technologies such as those that use high resolution, ion mobility, *etc.*, use this approach.[4] For immunoassays, an analogous approach would be the development of antibodies against the individual toxins. Obtaining antibodies that recognise only one congener (*e.g.* AFB_1) and not a closely related congener (*e.g.* AFB_2) can be difficult. This fact, as well as the ability of advancing MS techniques to detect potentially thousands of analytes (whether targeted or not), suggests that it will be difficult for antibodies to match the selectivity of MS for measuring individual toxins. However, the very cross-reactivity (CR) that is a disadvantage of antibodies for measuring individual toxins can be an advantage in certain circumstances or for certain types of assays. For example, antibody-based assays have the potential to be used to give an "integrated" or "summed" response. Consider the idealised situation where an antibody has identical CR to AFB_1 and AFB_2. In such a situation, the assay would detect the sum of $AFB_1 + AFB_2$. If it is the sum, and not the identity of the contributing toxins, that is important, then this is advantageous. The summed response may be desired in situations where a group of toxins, rather than the individual constituent toxins, is regulated. For example, the European Commission has established maximum levels of the sum of aflatoxins B_1, B_2, G_1 and G_2, as well as for AFB_1 alone.[5] The summed response may also be desirable in situations where the goal of the assay is to mimic the response from a toxicity assay. An example, in the case of marine toxins, is to mimic the mouse bioassay. Lastly, having an assay that yields a summed response may avoid additional sample preparation, such as the digestion or hydrolysis of cereal samples before measuring toxin content. There is a long history of the development of "generic" antibodies for recognising whole groups of mycotoxins; for example, the type A trichothecenes.[6]

Other potential advantages of immunoassays are speed and cost. Advances in MS and chromatographic technologies have rapidly reduced analysis times for such methods. It is not unusual for the final, determinative step of modern UHPLC-MS or UHPLC-fluorescence methods to be 5 minutes or less. For example, four aflatoxins can be chromatographed in less than 4 minutes.[7] While this is remarkable, in many cases such speed at the determinative step comes at the expense of the need for a more thorough and labour-intensive sample preparation and clean-up.

Immunoassays, which rely upon the selectivity of the antibody–toxin interaction, can reduce the need for extensive sample preparation and clean-up. For this reason, many of the commercial immunoassays can provide a complete analytical method that is rapid at all stages from sample preparation to the determinative step. The antibody-based and instrumentation-based approaches need not be mutually exclusive. Many current methods combine aspects of both technologies. The high selectivity of a modern mass spectrometer can be augmented through the use of antibody-based sample clean-up technologies such as immunoaffinity columns (IAC). Such columns can be used to isolate the toxins of interest, selectively removing impurities that might influence the ionisation process in the mass spectrometer. Hence, the issues of antibody interactions with the masked forms of mycotoxins may impact the advanced instrumental techniques as well.

3.1.3   CR of Immunoassays

Mycotoxins, as low-molecular-weight compounds, must generally be conjugated to a larger molecule (typically a protein, termed a carrier protein) in order to generate an immune response in animals. How this linkage is achieved can influence the performance characteristics of the resulting antibodies and, in particular, which types of compounds they will bind. The ability of an antibody binding site to accommodate structures other than the structure used during the antibody development process is termed CR. The concept of CR is central to understanding immunoassays to mycotoxin congeners and to determining whether such assays are useful for detecting masked mycotoxins. CR can be expressed several ways, most commonly as a ratio of the midpoints (IC_{50}) of calibration curves for the target analyte and the congener: That is:

$$CR\ (\%) = ([IC_{50}\ \text{of target analyte}]/[IC_{50}\ \text{of congener}]) \times 100 \qquad (3.1)$$

A CR of greater than 100% implies that the assay detects the congener better than the target analyte, while a CR of less than 100% indicates that it is less effective at detecting the congener than the target analyte. The ideal situation, where an immunoassay has a CR of 100% for all the congeners of interest, is unlikely. Good immunoassays are designed to detect primarily the target analyte, to detect other congeners of known interest and to have minimal cross-reaction with unrelated sample or matrix constituents.

CR is not confined to known compounds, so existing immunoassays may be able to detect materials that they were not originally designed to detect, including masked mycotoxins. CR of assays to congeners of the target (parent) mycotoxin is commonplace, although not an inherent property of all mycotoxin tests. Why does CR to related compounds exist? In part, this can be explained by the structural similarities between the congener and the parent toxin. It is logical that molecular interactions such as hydrogen bonds or Van der Waals forces between the antibody binding site and the parent toxin also occur between the binding site and equivalent functional groups on the congener. However, conjugates like the glucosides of deoxynivalenol (DON) or T-2 toxin (T2) have been significantly modified and include a large, bulky addition such as a glucose or oligosaccharide. Given the selectivity seen with good antibodies, binding of toxins modified with such groups might seem unlikely. The fact that it occurs suggests that there may be an additional aspect that has a bearing upon the binding interaction, namely the structure of the conjugate used for the immunisation (the immunogen). Because the target toxin has been modified by linking it to a carrier protein, it should not be surprising that some of the resulting antibodies recognise not only the toxin, but also accommodate structures similarly linked to carbohydrates, amino acids or other conjugated mycotoxins.

For trichothecenes it is common to produce protein conjugates for immunisation by linking through primary or secondary hydroxyls (Figure 3.1). As a result, antibodies made against DON linked through the 3- or 15-hydroxyls, for example, also generally recognise one or more of the acetylated congeners (*i.e.* 3-Ac-DON or 15-Ac-DON).[8–12] These are the same

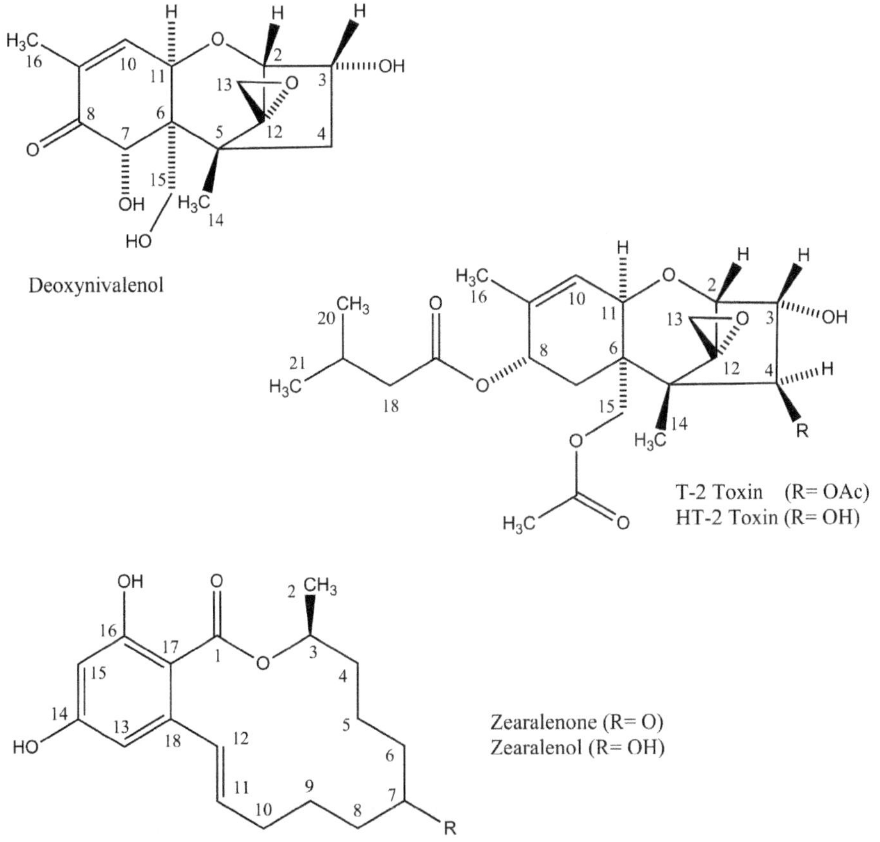

Figure 3.1 Structures of deoxynivalenol (DON), T-2 toxin (T2) and zearalenone (ZEN).

sites onto which glucose (or glucuronides in the case of mammalian de-toxification) is introduced. For zearalenone (ZEN), the most common route to producing antibodies has been to introduce a carboxymethyloxime (CMO) linkage through the 7-carbonyl position.[13] The selection of ZEN-specific antibodies that recognise free ZEN favours those that have good recognition of the region of the molecule at sites away from the linker. The glucosides or sulphate derivatives of ZEN are located on the C-14 and C-16 hydroxyl groups that are distal from the site of linkage (*i.e.* C-7; Figure 3.1).[14] This is in effect placing bulky groups directly onto sites where the antibody and ZEN would interact. As such, it would not be unusual for antibodies that have been selected to recognise ZEN (and therefore recognise the free hydroxyls at these positions) to not bind the masked forms conjugated through these sites.

Whether CR is beneficial or not depends upon the targets selected by the analyst. An assumption is often made that CR can lead to overestimation, and this can be true. An example would be a DON assay that has a 600% CR for 3-Ac-DON. In the idealised case, if the 3-Ac-DON were present at $0.17\,\mathrm{mg\,kg^{-1}}$ and DON were present at $1\,\mathrm{mg\,kg^{-1}}$, the assay would return a result closer to $2\,\mathrm{mg\,kg^{-1}}$, assuming that DON standards were used to calibrate the assay. However, what tends to be overlooked is that CR can also lead to underestimation. If there were an aflatoxin assay with a CR of 100% for AFB_1 and 50% for AFG_1, then if $10\ \mu\mathrm{g\,kg^{-1}}$ of each were actually present, the assay would return a value closer to $15\ \mu\mathrm{g\,kg^{-1}}$. Makers of mycotoxin test kits attempt to ensure that the responses of kits accommodate the types of congeners expected. Unfortunately, the pattern of toxins present may not mimic what has been anticipated. Until recently, the masked mycotoxins have fallen into the category of unanticipated analytes. However, within the past few years, efforts have been made to remedy this situation.

3.2 Immunoassays for Detecting Masked Mycotoxins

3.2.1 Assays Developed for Parent Mycotoxins that Cross-react with Masked Forms

3.2.1.1 *Immunoassays*

The ability of immunochemical methods, designed to detect the parent toxins, to recognise masked forms was recently reviewed.[1,2] An excellent summary of the CRs of commercial enzyme-linked immunosorbent assays (ELISAs) for masked mycotoxins was provided by Berthiller *et al.*[1] DON ELISAs from four commercial sources have been examined for CR to deoxynivalenol-3-glucoside (DON-3-Glc). These included ELISAs marketed by R-Biopharm (Darmstadt, Germany), Romer Labs (Tulln, Austria), EuroProxima (formerly Euro Diagnostica, Arnhem, The Netherlands) and

Table 3.1 Cross-reactivity of commercial DON immunoassays with 3-Ac-DON and DON-Glc.

Format	Kit name	Source	3-Ac-DON	DON-3-Glc	Matrix	Citation
ELISA	AgraQuant® 0.25/5.0	Romer Labs	392	45	Aq. solution	19
	AgraQuant® 0.25/5.0	Romer Labs	412	66	Spiked beer	19
	AgraQuant®	Romer Labs	770	52	Water	22
	DON EIA	Euro Diagnostica[a]	94	37	Aq. solution	19
	DON EIA	Euro Diagnostica[a]	60	53	Spiked beer	19
	DON EIA	Euro Proxima	230	115	Water	22
	Veratox® 5/5	Neogen Corp.	103	32	Aq. solution	19
	Veratox® 5/5	Neogen Corp.	117[b]	51[b]	Spiked beer	19
	Veratox® 5/5	Neogen Corp.	40	157	Water	22
	Ridascreen®	R-Biopharm	328	78	Aq. solution	19
	Ridascreen®	R-Biopharm	275	99	Spiked beer	19
	Ridascreen®	R-Biopharm	NR[c]	82 and 98	NR[c]	20
LFD	ROSA LF-DONQ®	Charm Sciences	60	8	Water	22
FPIA	Mycontrol	Aokin AG	167	22	Water	22

[a]Now EuroProxima.
[b]Measured at (B/Bo) of 60%, equivalent to IC_{40}. B: signal from sample, Bo: maximum signal developed.
[c]NR: Not reported.

Neogen Corp. (Lansing, MI, USA) (Table 3.1).[15–18] Four ELISA kits, from the same manufacturers, were also used to screen for DON content in malts.[17] The malts were further analysed by liquid chromatography–tandem mass spectrometry (LC-MS/MS) to directly establish the DON-3-Glc and acetyl-DON contents. The kits reported higher levels of "DON" than observed by LC-MS/MS and, in some cases, the results by ELISA were substantially higher than could be accounted for through summation of the levels of DON, DON-3-Glc and the acetyl-DONs. It was noted that the difference was particularly distinct for malts processed at temperatures above 120 °C.

In addition to ELISAs, a commercial lateral flow device (LFD), ROSA LF-DONQ (Charm Sciences, Lawrence, MA, USA), and a commercial fluorescence polarisation immunoassay (FPIA) known as Mycontrol DON (Aokin AG, Berlin, Germany) were also examined for CR to DON-3-Glc.[18] Because these two kits reported results differently from the ELISAs (*i.e.* directly as readings of $\mu g\,kg^{-1}$ in the foodstuff, rather than absorbance), the authors of that report used a different method for calculating the CR. This involved determining the slopes of a correlation analysis between the equivalent DON contents in cereal and the reported responses (*i.e.* the reported DON contents). This analysis was done for DON and for the congener, so that:

$$CR\ (\%) = ([\text{slope obtained for congener}]/[\text{slope obtained for DON}]) \times 100$$
$$(3.2)$$

All four of the ELISAs, the LFD and the FPIA cross-reacted with 3-Ac-DON and DON-3-Glc (Table 3.1).[15,16,18] It is clear that many of the commercial test kits for DON have a significant CR for DON-3-Glc, although the extent of the CR can be quite variable between manufacturers, across assay platforms and with matrix. The number of studies is too small to draw many conclusions. However, it is interesting to note that sometimes widely different results were obtained from kits from the same manufacturer. This difference could result from many factors, including the matrix; however, it is possible that the kits themselves, even from the same manufacturer, may have been different when tested at different times and in different locations. Interestingly, with one exception (Veratox® DON, with standards in water), the CR of most of the kits was lower for DON-3-Glc than for 3-Ac-DON. Although linked through the same site on DON, the glucose residue is substantially larger, so this may be the result of steric effects. In the latter case of Veratox®, there was a striking difference between the CR for DON reported in water (157%) and in aqueous solution or spiked beer (32% and 51%, respectively). What is also striking is the change in CR of the DON-3-Glc relative to that for 3-Ac-DON, from less than the 3-Ac-DON (in aqueous solution or spiked beer) to greater than the 3-Ac-DON (in water). Zachariasova *et al.*[15] noted that, with all four test kits they examined, the DON-3-Glc CRs were higher in beer than in aqueous solutions. This aspect of the potential effects of matrix on CR warrants further investigation.

Despite many differences, it appears that all of the commercial DON immunoassays that have been tested show CR to DON-3-Glc. Several non-commercial platforms for detecting DON have also been tested for CR to DON-3-Glc. These include a surface plasmon resonance (SPR) sensor and an imaging SPR (iSPR) sensor.[19,20] In the first of these, the SPR biosensor was developed to detect nivalenol (NIV) as well as DON.[19] The sensor used immobilised DON–bovine serum albumin (BSA), with competition between the DON–BSA and DON (or its congeners) for a limited amount of DON antibody. Interestingly, the sensor, while it showed good CR for NIV (52%), also showed almost equivalent CR for DON-3-Glc (60%). The second SPR device was based on a microarray format, with the assay constructed to detect DON and ZEN concurrently.[20] Using a commercially available antibody (Aokin), the measured CR to DON-3-Glc was 36%, while CRs for 3-Ac-DON and 15-Ac-DON were 71% and 66%, respectively. Incidentally, the CR for DON-3-Glc in this format was similar to that reported with the Mycontrol DON FPIA (22%).[18]

The literature on the CR of fumonisin immunoassays towards masked fumonisins is more limited. Early work established the potential for hidden fumonisins in cornflakes, which were suggested to be protein bound.[21]

In that context, "hidden" referred to fumonisins that were not detected following an extraction with a mixture of acetonitrile, methanol and water, but which could be detected by extracting the residue with a sodium dodecyl sulphate solution, followed by alkaline hydrolysis. Later, using a LC-MS/MS technique, it was further determined that significant amounts of fumonisins may exist in bound form in maize and maize-based products.[22] In that context, "bound" referred to derivatives of fumonisin (covalent or non-covalent) that were not detected without alkaline hydrolysis. Alkaline hydrolysis functioned both to render more of the fumonisin extractable, but also to remove the tricarballylic acid side chains from the fumonisins, yielding the corresponding aminopentol derivatives (hydrolysed fumonisins or HFBs). The response of the Ridascreen[®] Fumonisin ELISA to several fumonisin derivatives was evaluated and presented qualitatively.[22] As expected, the ELISA responded well to the intact FB_1 and poorly to HFB_1. Additional congeners that were tested included N-deoxyfructosyl-FB_1 (NDF-FB_1) and N-deoxy-fructosyl-HFB_1 (NDF-HFB_1), which are products of the reaction with glucose, the mono- and di-lysine derivatives of FB_1 (Lysil-FB_1), an FB_1–starch fraction and a protein (prolamin) fraction from a naturally contaminated sample. Of these, the NDF-FB_1, Lysil-FB_1, FB_1–starch and prolamin fractions all gave positive ELISA responses. Responses from the Lysil-FB_1 and prolamin fractions were indicated as being greater than for the free FB_1. LC-MS/MS data in the same article suggested that the globulin fraction of maize may also contain fumonisins not released during traditional extraction.[22] These results helped explain what heretofore had been suggested to be an overestimation of fumonisin content by immunoassays.

ZEN is commonly reduced at either the 11–12 double bond or the 7-carbonyl. Reduction of the double bond yields zearalanone, while reduction at the carbonyl yields α- or β-zearalenols. Reduction at both positions yields α- or β-zearalanols (Figure 3.1). The CR of ZEN immunoassays to these congeners has been widely described. However, there are few reports on CR with the masked forms of ZEN, and there are no reports on how commercial ELISAs or LFDs might detect these masked forms. One approach to measuring zearalenone-14-glucoside (ZEN-14-Glc) has been through hydrolysis to ZEN. In this case, the presence of the conjugate has been estimated by calculating the difference between the amounts found before and after cleavage.[23] Cleavage was obtained either by hydrolysis with trifluoromethanesulfonic acid or enzymatically using glycoside hydrolases, of which seven were tested. Immunoassays based upon five ZEN monoclonal antibodies (mAbs) were examined. Of these, four had specificity for ZEN and had CR of less than 10% for ZEN-14-Glc (referenced using an alternative nomenclature as ZEN-4-Glc in the article).[23] For one of these (mAb 5), the antibody was reportedly developed using an immunogen composed of ZEN-CMO linked to BSA. This linkage, through the 7-carbonyl of ZEN is on the opposite side of the ZEN to where the glucoside is located, which may be the reason for the low CR for the glucoside. Lastly for ZEN, there is a report in which a commercially available antibody (from Aokin) was applied in an iSPR format. In that format, the device did not cross-react with zearalenone-14-sulphate (ZEN-14-S).[20]

3.2.1.2   *Immunoaffinity Columns*

In order to improve the detection of chromatographic assays, IACs are often used to isolate selected toxins from food matrix. Because of this, the CR of the IAC for masked mycotoxins is an important determinant to whether the final assay can quantitatively detect the masked forms. Most laboratories do not make their own IACs and instead rely upon commercial sources, which may be constructed with antibodies that have widely different CR profiles. Two reports that examined the CR of DON IACs were those of Vendl *et al.*[24] and Versilovskis *et al.*[25] For DON, five products were tested in total. Two of these were from R-Biopharm (DONPREP® and DZT MS-Prep®) and one each were from Vicam (DONtest®), Neogen Corp. (NeoColumn®) and Aokin (ImmunoClean®). Neither the DONPREP® nor the DONtest® were able to recover DON-3-Glc from spiked maize.[24] The ImmunoClean® DON also had no reported CR to DON-3-Glc.[25] Interestingly, the FPIA for DON from the same manufacturer as the ImmunoClean® DON (Aokin) showed modest CR to DON-3-Glc (22%),[18] suggesting that either the manufacturer used different antibodies for the two formats or the format itself affected CR. Also, while in one report the DONPREP® was not be able to recover DON-3-Glc, in another report 58% was recovered, so there is some inconsistency in the literature.[24,25] The reason for the discrepancy is unclear, although it is worth noting that the two reports were testing very different matrices: maize and calf serum. With the spiked calf serum, the DONPREP®, DZT MS-Prep® and NeoColumn® also exhibited significant CR for the 3-glucuronide of DON, whereas the ImmunoClean® DON did not. That is, similar patterns of CR were seen with DON-glucuronide as were seen with the DON-3-Glc. There is also an indication that certain IACs directed against DON may also bind oligoglycosylated forms of DON.[26] In that report, the presence of DON and its glycosylated forms was investigated at the various stages of malt/beer production and bread making, and in naturally contaminated samples. Two types of DON IAC were used, in very different ways. The DONtest® HPLC IAC, demonstrated to be "non-cross-reacting" to the glucosides, was used to remove DON from malt extracts so as to separate it from DON-glucosides. The preparation, containing less DON, was then applied to a "cross-reactive" IAC (*i.e.* the DONPREP®). Recovery of DON-3-Glc from the DONPREP® IAC ranged from 4 to 102% depending upon whether and how much DON was also present to compete for available binding sites on the column. Malt extracts that did not bind to the DONtest®, but which did bind to the DONPREP® were eluted and then examined for the presence of the various glucosides by LC–high-resolution MS and were also subjected to enzymatic treatment. The process was used to indicate the presence of di-, tri- and tetra-glycosidic derivatives of DON in a beer sample.[26] The occurrence of DON, DON-3-Glc and the di- and tri-glucosides in naturally contaminated samples of malt, beer and commercial baked goods was demonstrated. This research represents an excellent example of using antibody selectivity to isolate specific masked mycotoxins.

As discussed earlier, the selectivity of the antibody depends in no small part upon how (and through what positions) the toxin was linked to the protein when the immunogen was made. Because many DON antibodies have been reportedly produced by linking through the C-3 hydroxyl, it is not surprising that activity towards DON-3-Glc is common. While beyond the scope of this review, it would be of interest to determine if those antibodies that were produced from immunogens linked through the C-15 hydroxyl recognise DON-3-Glc poorly in relation to those antibodies that were produced from immunogens linked through the C-3 hydroxyl, in addition to whether such antibodies might preferentially recognise a DON-15-Glc. Certain of the DON IACs also recognise related metabolites such as the de-epoxy DON (DOM-1). This has been used to detect DON or DOM-1 liberated from serum samples treated with β-glucuronidase.[27] Such enzymatic treatment to hydrolyse DON from DON-glucuronides has also been used with urine to help estimate human exposures.[28] Recently, the CR of commercial DON IAC for DON-3-Glc has been used to study the presence of this conjugate during the industrial process of making wholegrain crackers.[29] Estimates of masked DON have also been obtained by measuring the signal from ELISAs conducted before and after hydrolysis of corn or wheat samples with trifluoromethanesulfonic acid.[30]

Knowing the CR of the antibodies used in assays can help provide insights into previous observations in the literature. There is a report comparing the performance of the Ridascreen® ELISA, a non-commercial LFD, and an LC-MS/MS method for detection of DON in field inoculated wheat samples.[31] In that report, there was a good correlation between the LFD, which was a qualitative device, and the LC-MS/MS. There was a poor correlation between the ELISA and the LC-MS/MS. These results could, perhaps, be explained by examining the antibodies used in the two immunoassays. The LFD used an antibody with high CR for 15-Ac-DON and low CR for 3-Ac-DON. Given that the CR observed with most commercial tests for DON-3-Glc is less than that for 3-Ac-DON, it is not unreasonable to speculate that the CR of this assay for DON-3-Glc would probably be low. As noted earlier (Table 3.1), the Ridascreen® ELISA has a good CR with both 3-Ac-DON and DON-3-Glc. Therefore, it might be expected that samples containing significant levels of DON-3-Glc would report higher levels of "DON" with the Ridascreen® test than those reported with the LFD or the LC-MS/MS, which were able to distinguish DON separately.

While much of the work for masked trichothecenes has been with DON, recently several glucoside derivatives of T2 and HT-2 toxin (HT2) have been described.[32–35] Veprikova *et al.*[35] examined four extraction/clean-up procedures used to prepare samples for UHPLC-QqTOF (quadrupole-quadrupole time of flight mass spectrometry) analysis of T2, HT2 and their glucosides in naturally contaminated barley, wheat and oats. One of these entailed the use of the Easi-Extract[®] T2 and HT2 IACs (R-Biopharm). The IAC had sufficient CR when it was used to isolate T2, HT2, T2-Glc, HT2-Glc and the diglucoside of HT2 (HT-2-diGlc) from naturally contaminated samples. The CR was not quantified, perhaps because the IAC is less amenable than ELISA to this type of calculation.

For ZEN, six products were tested, five from the same companies that produced the DON IACs (above), and a sixth from Romer Labs. None of the products, which included Zearatest[®], ZearaStar[®], Easi-Extract[®] ZEN, DZT MS-Prep[®], NeoColumn[®] ZEN and ImmunoClean[®] C ZEN, cross-reacted significantly with ZEN-14-Glc.[24,25] The Easi-Extract[®] ZEN, DZT MS-Prep[®], NeoColumn[®] ZEN and ImmunoClean[®] C ZEN also did not allow recovery of ZEN-diglucoside, ZEN-16-Glc or ZEN-14-glucuronide.[25] In addition, the Easi-Extract[®] ZEN, Zearatest[®] and ZearaStar[®] did not allow recovery of α-zearalenol-14-β-D-glucopyranoside, β-zearalenol-14-β-D-glucopyranoside or ZEN-14-S from spiked maize.[24] These results may be related to the relative locations of the modifications to the ZEN backbone and the site of attachment of the toxin to the protein during the antibody development. As discussed earlier, antibodies with high affinities for ZEN that were produced using a CMO linkage through the C-7 carbonyl might not be able to accommodate the very different glucosides or glucuronides modified at the C-14 and C-16 positions. Interestingly, the Zearatest[®], ZearaStar[®] and Easi-Extract[®] ZEN all had significant CR to congeners of ZEN reduced at the C-7 position (*i.e.* the α- and β-zearalenols).[24] This pattern would support the speculation that the poor recognition of ZEN-14-Glc is derived from the use of immunogens linked through C-7 of ZEN, the most common site where ZEN–protein linkages have been reported in the literature.

3.2.2 Assays Developed Specifically for the Masked Forms

While it is clear that immunoassays developed against the parent toxin may recognise the corresponding masked forms, it is also clear that in many cases they do not. The contrast between the CR of DON assays towards DON-3-Glc and the CR of ZEN assays towards ZEN-14-Glc is the best example. This suggests that detecting certain of the masked mycotoxins will require efforts specifically directed at producing antibodies towards them, or at the very least efforts to change immunogens to more closely match (at site and linkage) those seen with the masked derivatives. As recently as in early 2013, there were no published immunoassays that had been specifically developed for masked mycotoxins.[1,2] At the time this chapter was written (March 2014), there was only one report in this area; however, efforts in this area are expected to increase rapidly due to the considerable interest in the detection of masked mycotoxins.

Recently, the 3-glucoside derivative of T2 (T2-3-Glc) and the 3- and 4-derivatives of HT2 (HT2-3-Glc, HT2-4-Glc) were discovered, and their presence was demonstrated in fungal cultures and naturally contaminated samples of wheat, oats and maize.[32–34] Relatively large amounts of T2-3-Glc were produced by feeding the yeast *Blastobotrys muscicola* with high levels of

T2.[36] Interestingly, the resulting form of the T2-3-Glc was an anomer with the glucose in the α configuration, rather than the β configuration observed with DON-3-Glc. The T2-3-Glc was conjugated to keyhole limpet hemocyanin and used to immunise mice, from which ten mAbs were selected.[37] Of the ten, six exhibited CR towards T2 ranging from 87 to 101% (*i.e.* similar to that of the T2-3-Glc) and two demonstrated greater CR towards T2 than the T2-3-Glc (CRs of 122 and 135%). The results suggested that obtaining antibodies that recognise both the parent toxin and the masked form was possible by using immunogens made with the masked form, opening this up as an avenue for producing antibodies capable of detecting both forms.

The importance of CR for the utility of the immunoassay has been described in the previous sections. However, an underappreciated aspect of CR is the degree to which it can change with solvent strength. This is because solvent strength influences not only the interaction between the target analyte and the antibody, but also the interaction between the masked form and the antibody. If those interactions do not change equally, then the CR of the masked form will change. As an example, consider the reported effects of solvent upon the IC_{50} for T2 and T2-3-Glc in ELISA.[37] One of the antibodies in that report (2–13) was evaluated at five concentrations of methanol and five concentrations of acetonitrile. To compare the effect of solvent, the authors used a statistic similar to CR, the relative response (RR):

$$RR\ (\%) = ([IC_{50}\ \text{of analyte in buffer}]/[IC_{50}\ \text{of analyte in solvent mixture}]) \times 100$$

$$(3.3)$$

Although very similar, RR is not the same as CR because it compares one analyte in two environments rather than two analytes in one environment. For T2-Glc in 20% methanol, the RR was 81%, whereas for 20% acetonitrile, the RR was 44%, indicating a greater influence of acetonitrile on assay performance.

Many aspects can influence the shape, as well as the IC_{50}, of immunoassay calibration curves. Therefore, while RR or CR are excellent statistics for comparing congeners or environmental effects, they may not fully capture the impact upon other assay parameters, such as limit of detection, limit of quantitation or detection range. For example, in buffer, the T2-3-Glc ELISA showed a very similar IC_{50} for T2-3-Glc as for T2. In this case, the IC_{50}s were very similar, so the CR of T2 relative to T2-3-Glc was near to ideal (91.6%).[37] With the toxins in 20% methanol, both IC_{50}s changed. This changed not only the CR (to 63%), but also the relative shapes of the calibration curves (Figure 3.2), with the T2-3-Glc exhibiting a steeper slope. Thus, while CR and RR are important statistics, it should not be forgotten that there are instances in which they do not fully capture the changes that can occur between congeners, or for a given toxin with alterations to the environment.

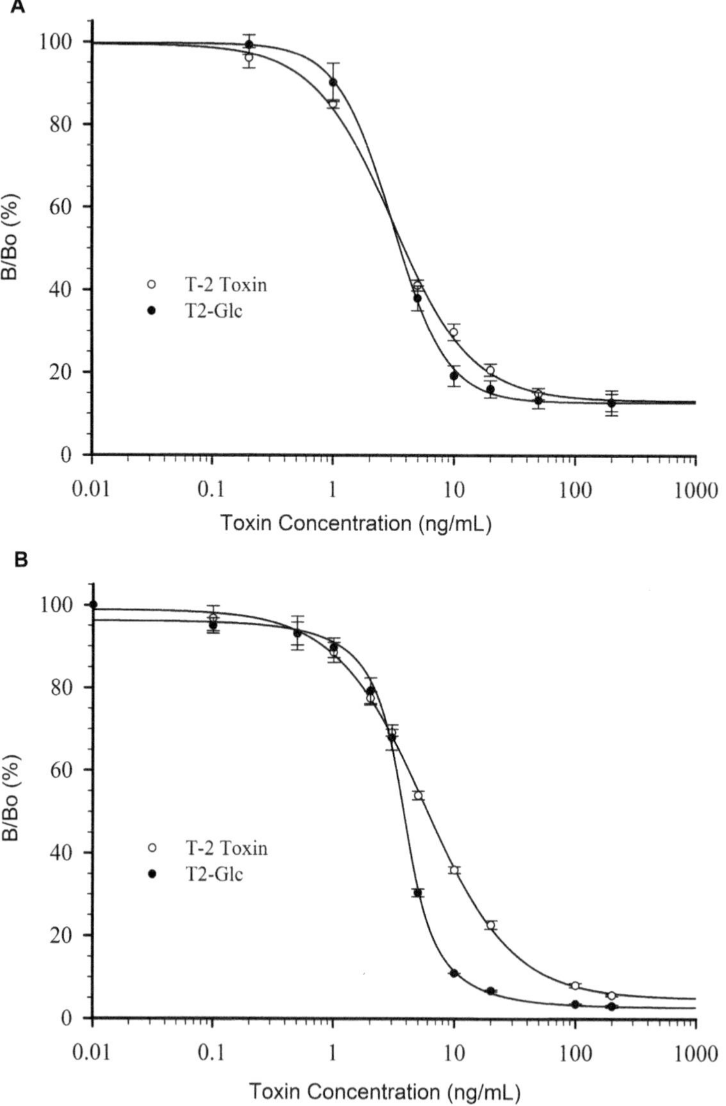

Figure 3.2   Effect of solvent upon cross-reactivity. Competitive indirect (CI)-ELISA results of T-2 toxin and its glucoside (T2-Glc) in (A) buffer and (B) 20% methanol. (B) is reproduced from *Toxins*, 2013, **5**, 1299–1313. The error bars are plus and minus one standard deviation from the mean.

3.3   Conclusion

Antibody-based methods for detecting and isolating masked mycotoxins have an inherent advantage relative to other techniques that are rooted in the concept of CR. Although contradictory, this is also a potential dis-advantage. The paradox of CR is universal, regardless of the particular "masked" or "parent" toxin combination. Many of the immunoassays that have been developed against DON are able to detect DON-3-Glc, which is an advantage if this analyte is deemed of interest and a disadvantage if it is not. Fortunately, in the latter case, there exist immunoassays that detect DON but not the DON-3-Glc. Achieving the desired CR is based upon the chemistry used to produce the immunogens that are administered to the animals. The site where plants or animals introduce polar groups as part of metabolism or detoxification may be different from the site that chemists use to produce the immunogens for antibody production. As a result, certain groups of masked mycotoxins are detected better than others with existing immunoassays. For example, while the CR of DON-3-Glc is common for DON assays, the CR of ZEN-Glc or ZEN-S is uncommon for ZEN IACs. The difference is likely based upon the proximity of the site (and type) of linkage present in the immunogen, relative to the site (and type) of linkage present in the masked derivative. The ability to tailor the assay response to the desired analytes by changing the antibodies upon which the assays are based is an important aspect of assay design. Because of this, detection of certain masked mycotoxins can be readily accomplished, if desired, by selecting an appropriately cross-reacting commercially available immunoassay. However, for certain of the masked mycotoxins, it is apparent that changing the underlying chemistry used to produce the immunogens will be necessary in order to achieve CR to the corresponding masked forms. For this reason, it should be expected that the development of immunoassays for masked mycotoxins will likely continue.

While the focus of this review has been on immunoassays, recent advances in molecular biology and chemistry have led to the development of a variety of toxin-binding materials with the potential to serve in places where antibodies have traditionally been used. Examples include antibody fragments and single-chain antibodies such as nanobodies,[38] binding materials based upon short peptides,[39] materials based upon nucleotides such as aptamers[40] and fully synthetic polymeric materials such as the molecularly imprinted polymers.[41] Readers should recognise that in some cases such materials may ultimately perform better than antibodies in certain applications. The author is unaware of the explicit application of one of these novel binding materials to masked mycotoxins, but given the interest in such toxins and the development of novel binding materials, this would seem to be a temporary situation.

There is significant interest in determining the prevalence of masked mycotoxins to facilitate risk assessment. Immunoassays are rapid, cost-effective screening tools. Of the many mycotoxins known to exist, few of their masked forms have been isolated and tested to demonstrate the presence or absence of cross-reaction. One reason for this has been the lack of analytical standards for the wide variety of masked forms. The exception has been DON-3-Glc, and this is perhaps the reason why much of the literature to date has focused on this derivative. The production of larger amounts of the masked mycotoxins would facilitate not only their testing in existing immunoassays, but also the development of immunoassays specific for their detection. The challenges of producing and isolating greater amounts of the masked mycotoxins are expected to be met, and with it the continued development of improved immunoassays for the masked mycotoxins.

Disclaimer

Mention of trade names or commercial products in this publication is solely for the purpose of providing specific information and does not imply recommendation or endorsement by the US Department of Agriculture (USDA). The USDA is an equal-opportunity provider and employer.

Acknowledgements

Figure 3.2B was reproduced from *Toxins*, 2013, **5**, 1299–1313, doi:10.3390/toxins5071299, under Creative Commons license (creativecommons.org/Licenses/by/3.0).

References

1. F. Berthiller, C. Crews, C. Dall'Asta, S. D. Saeger, G. Haesaert, P. Karlovsky, I. P. Oswald, W. Seefelder, G. Speijers and J. Stroka, Masked mycotoxins: A review, *Mol. Nutr. Food Res.*, 2013, **57**, 165–186.
2. I. Y. Goryacheva and S. De Saeger, Immunochemical detection of masked mycotoxins: A short review, *World Mycotoxin J.*, 2012, **5**, 281–287.
3. C. Dall'Asta, C. Falavigna, G. Galaverna, A. Dossena and R. Marchelli, *In Vitro* digestion assay for determination of hidden fumonisins in maize, *J. Agric. Food Chem.*, 2010, **58**, 12042–12047.
4. P. Li, Z. Zhang, X. Hu and Q. Zhang, Advanced hyphenated chromatographic-mass spectrometry in mycotoxin determination:current status and prospects, *Mass Spectrom. Rev.*, 2014, **32**, 420–452.
5. European Commission, *Off. J. Eur.* Union, Commission Regulation (EU) No. 165/2010. L.50, 8–12.
6. R. D. Wei and F. S. Chu, Production and characterization of a generic antibody against group A trichothecenes, *Anal. Biochem.*, 1987, **160**, 399–408.
7. M. E. Benvenuti and J. A. Burgess, Rapid analysis of aflatoxins in corn, cereals, and almonds using ACQUITY UPLC H-class system with fluorescence detection. Waters Corporation Application Note APNT10172781, 2010.
8. E. N. C. Mills, S. M. Alcock, H. A. Lee and M. R. A. Morgan, An enzyme-linked immunosorbent assay for deoxynivalenol in wheat, utilizing novel hapten derivatization procdures, *Food Agric. Immunol.*, 1990, **2**, 109–118.
9. E. Usleber, E. Märtlbauer, R. Dietrich and G. Terplan, Direct enzyme-linked immunosorbent assays for the detection of the 8-keto-trichothecene mycotoxins deoxynivalenol, 3-acetyldeoxynivalenol, and

15-acetyldeoxynivalenol in buffer solutions, *J. Agric. Food Chem.*, 1991, **39**, 2091–2095.

10. M. J. Nicol, D. R. Lauren, C. O. Miles and W. T. Jones, Production of a monoclonal antibody with specificity for deoxynivalenol, 3-acetyldeoxynivalenol and 15-acetyldeoxynivalenol, *Food Agric. Immunol.*, 1993, **5**, 199–209.

11. R. C. Sinha, M. E. Savard and R. Lau, Production of monoclonal antibodies for the specific detection of deoxynivalenol and 15-acetyldeoxynivalenol by ELISA, *J. Agric. Food Chem.*, 1995, **43**, 1740–1744.

12. C. M. Maragos and S. P. McCormick, Monoclonal antibodies for the mycotoxins deoxynivalenol and 3-acetyl-deoxynivalenol, *Food Agric. Immunol.*, 2000, **12**, 181–192.

13. D. Thouvenot and R. F. Morfin, Radioimmunoassay for zearalenone and zearalanol in human serum: production, properties, and use of porcine antibodies, *Appl. Environ. Microbiol.*, 1983, **45**, 16–23.

14. M. P. Kovalsky Paris, W. Schweiger, C. Hametner, R. Stückler, G. J. Muehlbauer, E. Varga, R. Krska, F. Berthiller and G. Adam, Zearalenone-16-O-glucoside: a new masked mycotoxin, *J. Agric. Food Chem.*, 2014, **62**, 1181–1189.

15. M. Zachariasova, J. Hajslova, M. Kostelanska, J. Poustka, A. Krplova, P. Cuhra and I. Hochel, Deoxynivalenol and its conjugates in beer: a critical assessment of data obtained by enzyme-linked immunosorbent assay and liquid chromatography coupled to tandem mass spectrometry, *Anal. Chim. Acta*, 2008, **625**, 77–86.

16. J. Ruprich and V. Ostry, Immunochemical methods in health risk assessment: Cross reactivity of antibodies against mycotoxin deoxynivalenol with deoxynivalenol-3-glucoside, *Cent. Eur. J. Public Health*, 2008, **16**, 34–37.

17. M. Kostelanska, J. Hajslova, M. Zachariasova, A. Malachova, K. Kalachova, J. Poustka, J. Fiala, P. M. Scott, F. Berthiller and R. Krska, Occurrence of deoxynivalenol and its major conjugate, deoxynivalenol-3-glucoside, in beer and some brewing intermediates, *J. Agric. Food Chem.*, 2009, **57**, 3187–3194.

18. E. K. Tangni, J. C. Motte, A. Callebaut and L. Pussemier, Cross-reactivity of antibodies in some commercial deoxynivalenol test kits against some fusariotoxins, *J. Agric. Food Chem.*, 2010, **58**, 12625–12633.

19. T. Kadota, Y. Takezawa, S. Hirano, O. Tajima, C. M. Maragos, T. Nakajima, T. Tanaka, Y. Kamata and Y. Sugita-Konishi, Rapid detection of nivalenol and deoxynivalenol in wheat using surface Plasmon resonance immunoassay, *Anal. Chim. Acta*, 2010, **673**, 173–178.

20. D. Dorokhin, W. Haasnoot, M. C. R. Franssen, H. Zuilhof and M. W. F. Nielen, Imaging surface plasmon resonance for multiplex microassay sensing of mycotoxins, *Anal. Bioanal. Chem.*, 2011, **400**, 3005–3011.

21. E. K. Kim, P. M. Scott and B. P. Y. Lau, Hidden fumonisin in corn flakes, *Food Addit. Contam.*, 2003, **20**, 161–169.

22. C. Dall'Asta, G. Galaverna, G. Aureli, A. Dossena and R. Marchelli, A LC/ MS/MS method for the simultaneous quantification of free and masked fumonisins in maize and maize-based products, *World Mycotoxin J.*, 2008, **1**, 237–246.
23. N. V. Beloglazova, M. De Boevre, I. Y. Goryacheva, S. Werbrouck, Y. Guo and S. De Saeger, Immunochemical approach for zearalenone-4-glucoside determination, *Talanta*, 2013, **106**, 422–430.
24. O. Vendl, F. Berthiller, C. Crews and R. Krska, Simultaneous determination of deoxynivalenol, zearalenone, and their major masked metabolites in cereal-based food by LC-MS-MS, *Anal. Bioanal. Chem.*, 2009, **395**, 1347–1354.
25. A. Versilovskis, B. Huybrecht, E. K. Tangni, L. Pussemier, S. De Saeger and A. Callebaut, Cross-reactivity of some commercially available deoxynivalenol (DON) and zearalenone (ZEN) immunoaffinity columns to DON- and ZEN-conjugated forms and metabolites, *Food Addit. Contam., Part A*, 2011, **28**, 1687–1693.
26. M. Zachariasova, M. Vaclavikova, O. Lacina, L. Vaclavik and J. Hajslova, Deoxynivalenol oligoglycosides: new masked Fusarium toxins occurring in malt, beer, and breadstuff, *J. Agric. Food Chem.*, 2012, **60**, 9280–9291.
27. J. He, X. Z. Li and T. Zhou, Sample clean-up methods, immunoaffinity chromatography and solid phase extraction, for determination of deoxynivalenol and deepoxy deoxynivalenol in swine serum, *Mycotoxin Res.*, 2009, **25**, 89–94.
28. F. A. Meky, P. C. Turner, A. E. Ashcroft, J. D. Miller, Y.-L. Qiao, M. J. Roth and C. P. Wild, Development of a urinary biomarker of human exposure to deoxynivalenol, *Food Chem. Toxicol.*, 2003, **41**, 265–273.
29. M. Suman, A. Manzitti and D. Catellani, A design of experiments approach to studying deoxynivalenol and deoxynivalenol-3-glucoside evolution throughout industrial production of wholegrain crackers exploiting LC-MS/MS techniques, *World Mycotoxin J.*, 2012, **5**, 241–249.
30. S. T. Tran and T. K. Smith, Determination of optimal conditions for hydrolysis of conjugated deoxynivalenol in corn and wheat with trifluoromethanesulfonic acid, *Anim. Feed Sci. Technol.*, 2011, **163**, 84–92.
31. A. Y. Kolosova, L. Sibanda, F. Dumoulin, J. Lewis, E. Duveiller, C. van Peteghem and S. De Saeger, Lateral-flow colloidal gold-based immunoassay for the rapid detection of deoxynivalenol with two indicator ranges, *Anal. Chim. Acta*, 2008, **616**, 235–244.
32. M. Busman, S. M. Poling and C. M. Maragos, Observation of T-2 toxin and HT-2 toxin glucosides from *Fusarium sporotrichioides* by liquid chromatography coupled to tandem mass spectrometry (LC-MS/MS), *Toxins*, 2011, **3**, 1554–1568.
33. V. M. T. Lattanzio, A. Visconti, M. Haidukowski and M. Pascale, Identification and characterization of new *Fusarium* masked mycotoxins, T2 and HT2 glycosyl derivatives, in naturally contaminated wheat and oats by liquid chromatography-high-resolution mass spectrometry, *J. Mass Spectrom.*, 2012, **47**, 466–475.

34. H. Nakagawa, S. Sakamoto, Y. Sago, M. Kushiro and H. Nagashima, The use of LC-Orbitrap MS for the detection of *Fusarium* masked mycotoxins: the case of type A trichothecenes, *World Mycotoxin J.*, 2012, **5**, 271–280.
35. Z. Veprikova, M. Vaclavikova, O. Lacina, Z. Dzuman, M. Zachariasova and J. Hajslova, Occurrence of mono- and di-glycosylated conjugates of T-2 and HT-2 toxins in naturally contaminated cereals, *World Mycotoxin J.*, 2012, **5**, 231–240.
36. S. P. McCormick, N. P. J. Price and C. P. Kurtzman, Glucosylation and other biotransformations of T-2 toxin by yeasts of the *Trichomonascus* clade, *Appl. Environ. Microbiol.*, 2012, **78**, 8694–8702.
37. C. M. Maragos, C. Kurtzman, M. Busman, N. Price and S. McCormick, Development and evaluation of monoclonal antibodies for the glucoside of T-2 toxin (T2-Glc), *Toxins*, 2013, **5**, 1299–1313.
38. P. J. Doyle, M. Arbabi-Ghahroudi, N. Gaudette, G. Furzer, M. E. Savard, S. Gleddie, M. D. McLean, C. R. Mackenzie and J. C. Hall, Cloning, expression, and characterization of a single-domain antibody fragment with affinity for 15-acetyl-deoxynivalenol, *Mol. Immunol.*, 2008, **45**, 3703–3713.
39. G. Giraudi, L. Anfossi, C. Baggiani, C. Giovannoli and C. Tozzi, Solid-phase extraction of ochratoxin A from wine based on a binding hexapeptide prepared by combinatorial synthesis, *J. Chromatogr. A*, 2007, **1175**, 174–180.
40. J. Cruz-Toledo, M. Dumontier, M. McKeague, X. Zhang, A. Giamberardino, E. McConnell, T. Francis and M. C. DeRosa, Aptamer base: A collaborative knowledge base to describe aptamers and SELEX experiments. *Database*, 2012, Article ID bas006.
41. M. Cichna-Markl, New strategies in sample clean-up for mycotoxin analysis, *World Mycotoxin J.*, 2011, **4**, 203–215.

Untargeted Analysis of Modified Mycotoxins using High-resolution Mass Spectrometry

MARTHE DE BOEVRE,[a,b] EMMANUEL NJUMBE EDIAGE,[a] CHRISTOF VAN POUCKE[a] AND SARAH DE SAEGER*[a]

[a] Laboratory of Food Analysis, Department of Bioanalysis, Faculty of Pharmaceutical Sciences, Ghent University, 9000 Ghent, Belgium; [b] Laboratory of Brewery Technology, Department of Bio- and Food Sciences, Faculty of Science and Technology, University College Ghent, Ghent, Belgium
*Email: sarah.desaeger@ugent.be

4.1 Analysis of Modified Mycotoxins by Traditional Liquid Chromatography–tandem Mass Spectrometry Methods

To harmonize the terminology in scientific parlance, the term 'modified' mycotoxins has been recently introduced.[1] In order to encompass all possible forms in which mycotoxins and their modifications can occur, the proposition has been made to draft a systematic definition consisting of four hierarchic levels. The highest level differentiates the *free* forms of mycotoxins (*e.g.* deoxynivalenol [DON]) from those being *matrix-associated* (*e.g.* DON-oligosaccharides) and from those being *modified* in their

Issues in Toxicology No. 24
Masked Mycotoxins in Food: Formation, Occurrence and Toxicological Relevance
Edited by Chiara Dall'Asta and Franz Berthiller

chemical structure. The lower levels further differentiate modified mycotoxin into *biologically modified* (*e.g.* deoxynivalenol-3-*O*-glucoside [DON-3-Glc], deoxynivalenol-3-*O*-glucuronide [DON-3-GlcA] and zearalenone-14-*O*-sulfate [ZEN-14-S]) and *chemically modified*, dividing the latter into *thermally formed* (*e.g.* nor-deoxynivalenol A) and *non-thermally formed* (*e.g.* deoxynivalenol-sulfonate). The term 'masked mycotoxins' is kept for the fraction of biologically modified mycotoxins that are conjugated by plants. The terminology described by Rychlik *et al.* (2014) has been applied in this chapter.[1]

The main focus in analysis of modified mycotoxins is pointed towards liquid chromatography–tandem mass spectrometry (LC-MS/MS); however, gas chromatography (GC) methods are also at hand for the quantification of trichothecenes, zearalenone (ZEN), ochratoxin A (OTA), fumonisins and its modified forms.[2,3] While derivatization is needed to render the mycotoxins volatile, extractable glycosylated derivatives are far too polar for derivatization. For this reason, only methods for acetylated forms, 3-acetyl-deoxynivalenol (3-Ac-DON) and 15-acetyl-deoxynivalenol (15-Ac-DON), are described. These are derivatized by means of a trimethylsilylimidazole (TMS) reagent containing TMS-trimethylchlorosilane-ethyl acetate (10/90, v/v/v).[4] Tran and Smith (2011) developed an indirect determination of modified DON using GC-MS *via* hydrolysis with trifluoromethanesulfonic acid. However, the application was not suitable for identifying any modified forms.

LC-MS/MS, however, is a hyphenated detection technique, and is particularly useful for the simultaneous determination of multiple (modified) mycotoxins. Moreover, as only three derivatives (DON-3-Glc, 3-Ac-DON and 15-Ac-DON) are nowadays commercially available as reference standards to be used for monitoring purposes, the identification power of mass spectrometry (MS) detection may be fully exploited for the identification of unknown derivatives as well.[5]

LC-MS/MS offers a powerful tool for the identification and characterization of polar non-volatile mycotoxins and their modified forms. A wide series of MS-based multi-(modified)-mycotoxin methods was recently proposed as described in Table 4.1.[6,7] Berthiller *et al.* (2005) were the first to describe the occurrence of DON and its modified forms using QTrap® LC-MS/MS and atmospheric-pressure chemical ionization (APCI) (structural information) and ultraviolet detection (quantification).[8] The same authors, in 2006, used the same instrumentation for the analysis of modified ZEN forms; however, single-reaction monitoring was selected for the analysis, while for identification and characterization of the derivatives, enhanced product ion (MS/MS) and multiple MS^3 modalities were chosen.[9]

Sulyok *et al.* (2006) described a method for the determination of 39 free and modified mycotoxins in two consecutive chromatographic runs (electrospray negative and positive [ESI$^-$ and ESI$^+$]), while Vendl *et al.* (2009) delineated an LC-MS/MS method using both ESI$^-$ and APCI probes for the simultaneous determination of eight modified forms in one ESI$^-$ run.[10,11] The methods were performed using QTrap® LC-MS/MS, fully exploiting the

Table 4.1 Overview of modified *Fusarium* mycotoxin analysis using LRMS and HRMS.[a]

Mycotoxin	Matrix	LC equipment	Mobile phase	Column	MS equipment	Ionization mode	Mass range (m/z)	Ref.
Deoxynivalenol-3-O-glucoside	Beer	Finnigan Surveyor LC	5 mM NH_4-acetate H_2O/MeOH (gradient)	Gemini C18	Hybrid LTQ orbitrap XL	ESI$^+$	90–900	33
Deoxynivalenol-3-O-glucoside	Cereal-based food	Finnigan Surveyor LC	H_2O/MeOH + 0.5% acetic acid (gradient)	Kinetex C_{18}	Linear ion trap LXQ	ESI$^-$	SRM	34
Deoxynivalenol-3-O-glucoside, Deoxynivalenol-3-O-diglucoside, Deoxynivalenol-3-O-triglucoside, Deoxynivalenol-3-O-tetraglucoside	Beer	Accela UHPLC	2 mM NH_4-formate H_2O/ACN (gradient)	Acquity UPLC BEH amide	LC orbitrap MS Exactive	ESI$^-$/ESI$^+$	120–2000	35
Nivalenol-3-O-glucoside	Wheat	Agilent 1200 series	10 mM NH_4-acetate/ ACN (gradient)	InertSustain C18	Agilent 6530 QTOF	ESI$^-$	100–1000	12
Deoxynivalenol-sulfonate 1, 2 and 3	Animal feed	Agilent 1290 Infinity	H_2O/MeOH + 0.1% formic acid (gradient)	SB-C18	Agilent 6550 *i*Funnel QTOF	ESI$^-$	100–1000	15
DON-glutathione	Wheat	Accela UHPLC	H_2O/MeOH + 0.1% formic acid (gradient)	XBridge C18	LTQ orbitrap XL	ESI$^+$	100–1000	37
Fusarenon-X-3-O-glucoside	Maize	Accela UHPLC	10 mM NH_4-acetate H_2O/MeOH (gradient)	HyPurity C18	LC orbitrap MS Exactive	ESI$^+$	70–1000	36
Neosolaniol-3-O-glucoside	Maize	Accela UHPLC	10 mM NH_4-acetate H_2O/MeOH (gradient)	HyPurity C18	LC orbitrap MS Exactive	ESI$^+$	70–1000	36
Diacetoxyscirpenol-3-O-glucoside	Maize	Accela UHPLC	10 mM NH_4-acetate H_2O/MeOH (gradient)	HyPurity C18	LC orbitrap MS Exactive	ESI$^+$	70–1000	36

Monoacetoxyscirpenol-3-*O*-glucoside	Maize	Accela UHPLC	10 mM NH_4-acetate H_2O/MeOH (gradient)	HyPurity C18	LC orbitrap MS Exactive	ESI^+	70–1000	36
T2-3-*O*-glucoside, HT2-3-*O*-glucoside, HT2-3-*O*-diglucoside	Maize	Accela UHPLC	10 mM NH_4-acetate H_2O/MeOH (gradient)	HyPurity C18	LC orbitrap MS Exactive	ESI^+	70–1000	39
T2-3-*O*-triglucoside	Wheat	Accela UHPLC	10 mM NH_4-acetate H_2O/MeOH (gradient)	Zorbax Eclipse C18	LC orbitrap MS Exactive	ESI^+	80–1000	40
T2-3-*O*-glucoside, HT2-3-*O*-glucoside, HT2-4-*O*-glucoside	Wheat and oats	Accela UHPLC	1 mM NH_4-acetate + 0.5% acetic acid, H_2O/MeOH (gradient)	Gemini C18	LC orbitrap MS Exactive	ESI^+	50–1000	32
T2-3-*O*-glucoside, T2-3-*O*-diglucoside HT2-3-*O*-glucoside, HT2-3-*O*-diglucoside	Barley, wheat and oats	Dionex Ultimate 3000 UHPLC	5 mM NH_4-acetate H_2O/MeOH (gradient)	Acquity UPLC HSS T3	5600 TripleTOF	ESI^+	100–1000	42
Zearalenone-14-*O*-glucoside Zearalenone-16-*O*-glucoside	Barley, wheat and *Brachypodium distachyon* cultures	Agilent 1290 Infinity	H_2O/MeOH (30/70, v/v) (gradient)	Zorbax SB-C18 RRHD	Agilent 6550 *i*Funnel QTOF	ESI^-/ESI^+	50–1000	53
Modified zearalenone	*Arabidopsis thaliana*	Agilent 1100 series	5 mM NH_4-acetate H_2O/MeOH (gradient)	Aquasil® RP18	QTrap® MS/MS	ESI^-	SRM	49
Zearalenone-14-*O*-glucoside	Wheat	Waters 2690 Alliance	ACN/FA	Nucleosil® C18	VG® Quadrupole	ESI^+	SRM	50
Multi-modified	Cereal-based food	Agilent 11000 series	ACN/H_2O + 5 mM NH_4-acetate	Synergy® Polar RP18	4000 QTrap® MS/MS	ESI^-	SRM	51

[a]NH_4-acetate: Ammonium acetate; H_2O: Water; MeOH: Methanol; ACN: Acetonitrile; ESI^-: Electrospray in negative ionization mode; ESI^+: Electrospray in positive ionization mode; FA: Formic acid; HOAc: Acetic acid; SRM: Single-reaction monitoring; MS: Mass spectrometry; LC: Liquid chromatography; UHPLC: Ultra-high-performance liquid chromatrography; QTOF: Quadrupole time-of-flight; UPLC BEH: UPLC Ethylene Hybrid Bridged; UPLC HSS: UPLC High Strength Silica; RRHD: Rapid Resolution High Definition.

selectivity of MS detection and avoiding any clean-up steps in an over-simplified extract.

4.2 Analysis of Modified Mycotoxins and Possibilities of High-resolution Mass Spectrometry

Since the first reports on the use of LC-MS(/MS) for (multi-)mycotoxin analysis were published as described in the previous section, MS(/MS) has become a well-established technique for this matter. Compared to this, the use of high-resolution MS (HRMS) in mycotoxin analysis has been very recent; however, its popularity is increasing. Typical detectors used in HRMS are Fourier transform ion cyclotron resonance MS, magnetic sectors, orbitrap MS or time-of-flight MS (TOF-MS). In the last decade, TOF and orbitrap mass analyzers have been mainly chosen since these techniques are becoming more affordable and have demonstrated some advantages: identification, screening of non-target compounds and retrospective data analysis.[2,3]

Resolution in HRMS refers to the ability of the detector to separate two peaks of slightly different mass-to-charge ratios (m/z). Most commonly, resolution is expressed as $(m/z)/\Delta(m/z)$, where $\Delta(m/z)$ is the full width of the peak at half its maximum (FWHM) height.[4] Although the number of published applications on the use of HRMS in (modified) mycotoxin analysis is scarce when compared to low-resolution (multi-stage) MS, HRMS has several advantages over the latter.

One of the biggest advantages of HRMS is the ability to record full-scan spectra with measurement of the accurate mass of the analytes. The accurate mass is the experimentally determined exact mass of the molecule, while the exact mass is the sum of the most abundant isotopes of the constituent atoms of this molecule.[4] The more closely this accurate mass approaches the exact mass of the molecule (*i.e.* the more accurately the HRMS can measure the exact mass of the molecule), the better the elemental composition of the molecule can be postulated. The mass accuracy of a measurement is reported as the relative difference between the measured mass and the exact mass, expressed in parts per million (ppm). This accurate mass measurement allows for the detection of new compounds and the proposition of the elemental composition of these detected compounds.[5–8] Although this type of information can be used in the identification of new modified mycotoxins, it still remains difficult to use this matter to identify new compounds without any additional information. Supplementary information that can simplify this identification is the accurate mass data of the fragment ions. Although this information (MS and MS/MS data) can be obtained successively,[9–14] such data can also easily be obtained in the same analysis when using hybrid HRMS instruments such as quadrupole-orbitrap (Q-orbitrap) or Q-TOF. With such instruments, it is possible to record accurate masses of both precursor and fragment ions in one single run by using data-dependent analysis (DDA) or information data analysis (IDA) modes, where full-scan accurate mass fragmentation spectra are recorded when analytes are present in a sample above a predefined threshold.[15] Although the above described approach can provide the necessary accurate mass data for proposing an elemental composition and identifying the new compound, the large amount of data that is generated with such a DDA/IDA experiment makes it complicated to locate the new compound in the generated data. Indeed, the concentration of modified mycotoxins is often so low that no clear peak for this compound can be observed in the total ion chromatogram of either the full-scan MS or MS/MS data. To overcome this problem and to facilitate the detection of new unknown compounds, a metabolomics approach is often followed.[16,17] This implies that a number of biological and technical replicates of samples that are suspected to contain or to lack this new compound are analyzed with HRMS. After deconvolution of the acquired chromatograms and alignment to correct for differences in retention time between two runs, pattern recognition techniques, such as principal component analysis or linear discriminant analysis, are used to highlight the ions that are responsible for the differences between the groups of samples. Next, the accurate masses of these ions (and their fragments) can be used to postulate the elemental composition and/or to search online databases to identify this compound.

Besides the advantage in structural elucidation, the ability of HRMS to record full-scan spectra without loss in mass accuracy is also used in untargeted analysis or screening.[18] The capability of HRMS to record full-scan spectra results in a theoretically unlimited number of compounds that can be detected simultaneously at low concentration levels.[19] Consequently, HRMS can be used as a screening method to simultaneously detect a large number of compounds, often belonging to different classes, at the same time.[20–23] To include as many different compounds or classes of compounds in one single method, generic sample preparation procedures are often used. Such generic procedures, however, result in the ubiquitous presence of matrix compounds in the final extract, which can interfere with the detection, quantification and identification of the target compounds. Accurate mass measurement in HRMS allows interfering matrix compounds to be filtered out through the technique of mass clean-up. Due to the difference between the nominal mass of a compound (*i.e.* the sum of the integer masses of the most abundant isotopes in a molecule) and its exact or monoisotopic mass, a mass defect exists for each molecule.[4] By extracting the exact mass of the target analyte from the chromatogram, separation of co-eluted compounds or the target analyte from the matrix can be achieved. The higher the mass defect of the target analyte, because of chemical elements with an outstanding mass defect, the more efficient HRMS becomes for filtering out compounds from matrix interferences or co-eluting compounds.[19] Although HRMS in such a screening approach is used in an untargeted mode, the data processing is often done in a targeted way.[24,25] By having the software to explore for exact masses of a list of compounds in the full-scan spectra and additionally compare the DDA/IDA fragment spectra with reference spectra in a (self-constructed) library, one is actually looking for a limited number of compounds in an 'unlimited dataset'.[3] Also, because full-scan spectra were recorded, this HRMS screening approach allows retrospective analysis of the acquired full-scan data; consequently, the presence of 'newly discovered' compounds can be investigated in the data of prior-analyzed samples.

4.3 Untargeted Analysis of Modified *Fusarium* Mycotoxins in Natural Products

In the scope of this section, two major groups of mycotoxins produced by *Fusarium* spp. were investigated. Most dominant mycotoxin production includes trichothecenes and myco-estrogens. These are undoubtedly key substances in the (modified) mycotoxin incidence in cereals and cereal-based commodities, and have been the base for expanding (modified) mycotoxin research since 2009.[26] The following summary will provide the reader with an overview of new modified *Fusarium* mycotoxins discovered using HRMS.

4.3.1 *Fusarium* Mycotoxins: Trichothecenes

To date, more than 400 mycotoxins have been reported.[27] Trichothecenes are a family of naturally occurring tetracyclic sesquiterpenoids and in terms of their functional groups, they are divided into four groups (A, B, C and D). Most compounds contain an epoxide ring and a double bond at C_{12}–C_{13} and C_9–C_{10}, respectively (12,13-epoxytrichothec-9-ene) (Figure 4.1).[28] It is the epoxide ring, when present, that exerts the toxic activity of the trichothecenes.

Type A trichothecenes, mainly produced by *F. poae*, *F. langsethiae* and *F. sporotrichioides*, include the highly toxic T-2 toxin (T2), its deactylated form HT-2 toxin (HT2), diacetoxyscirpenol (DAS), monoacetoxyscirpenol (MAS) and neosolaniol (NEO).[29] Type B trichothecenes are distinguished from type A trichothecenes by the presence of a keto-group at C_8 and include the important trichothecenes nivalenol (NIV), DON, the acetylated forms 3-acetyldeoxynivalenol (3-Ac-DON) and 15-acetyldeoxynivalenol (15-Ac-DON) and fusarenon-X (FUSX). They are produced by *F. graminearum*, *F. equiseti* and *F. culmorum*.[30] Type C trichothecenes, like crotocin, contain an additional C_7–C_8 epoxide; however, they are not produced by *Fusarium* species. Type D trichothecenes are also non-*Fusarium* mycotoxins and contain a macrocyclic ring linking at C_4 and C_{15} with di- or tri-esters. These airborne *Stachybotrys* mycotoxins include satratoxins, roridins and verrucarins, and are prevalent in indoor environments.[31] Type A and B trichothecenes are widely distributed in cereals as natural pollutants, whereas the macrocyclic trichothecenes rarely occur in food and feed.

One of the most widely investigated masked forms of trichothecenes are the glucoside conjugates. The availability of the commercial reference

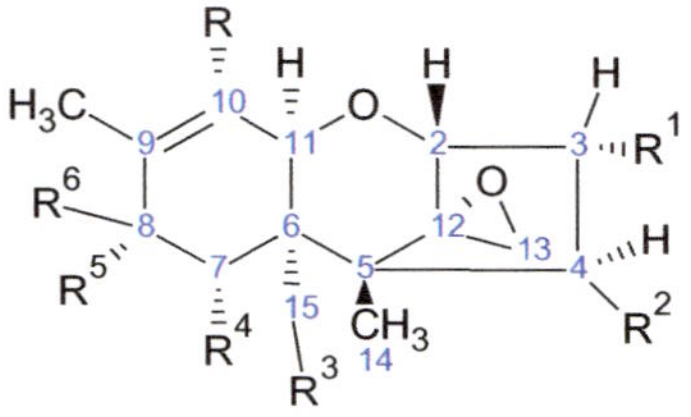

Type A trichothecenes

Modified mycotoxin	R^1	R^2	R^3	R^4	R^5	R^6	R
T-2 toxin-3-*O*-glucoside(T2-3-Glc)	$C_6H_{12}O_6$	$OCOCH_3$	$OCOCH_3$	H	$OCOCH_2CH(CH_3)_2$	H	H
T-2 toxin-3-*O*-diglucoside (T2-3-diGlc)	$C_{12}H_{24}O_{12}$	$OCOCH_3$	$OCOCH_3$	H	$OCOCH_2CH(CH_3)_2$	H	H
T-2 toxin-3-*O*-triglucoside (T2-3-triGlc)	$C_{18}H_{36}O_{18}$	$OCOCH_3$	$OCOCH_3$	H	$OCOCH_2CH(CH_3)_2$	H	H
HT-2 toxin-3-*O*-glucoside (HT2-3-Glc)	$C_6H_{12}O_6$	OH	$OCOCH_3$	H	$OCOCH_2CH(CH_3)_2$	H	H
HT-2 toxin-4-*O*-glucoside (HT2-4-Glc)	OH	$C_6H_{12}O_6$	$OCOCH_3$	H	$OCOCH_2CH(CH_3)_2$	H	H
HT-2 toxin-3-*O*-diglucoside (HT2-3-diGlc)	$C_{12}H_{24}O_{12}$	OH	$OCOCH_3$	H	$OCOCH_2CH(CH_3)_2$	H	H
Neosolaniol-3-*O*-glucoside (NEO-3-Glc)	$C_6H_{12}O_6$	$OCOCH_3$	$OCOCH_3$	H	OH	H	H
Diacetoxyscirpenol-3-*O*-glucoside (DAS-3-Glc)	$C_6H_{12}O_6$	$OCOCH_3$	$OCOCH_3$	H	H	H	H
Monoacetoxyscirpenol-3-*O*-glucoside (MAS-3-Glc)	$C_6H_{12}O_6$	$OCOCH_3$	H	H	H	H	H

Type B trichothecenes

Modified mycotoxin	R^1	R^2	R^3	R^4	R^5 and R^6	R
Deoxynivalenol-3-*O*-glucoside (DON-3-Glc)	$C_6H_{12}O_6$	H	OH	OH	O	H
Deoxynivalenol-3-*O*-diglucoside (DON-3-diGlc)	$C_{12}H_{24}O_{12}$	H	OH	OH	O	H
Deoxynivalenol-3-*O*-triglucoside (DON-3-triGlc)	$C_{18}H_{36}O_{18}$	H	OH	OH	O	H
Deoxynivalenol-3-*O*-tetraglucoside (DON-3-tetraGlc)	$C_{24}H_{48}O_{24}$	H	OH	OH	O	H
Nivalenol-3-*O*-glucoside (NIV-3-Glc)	$C_6H_{12}O_6$	OH	OH	OH	O	H
Deoxynivalenol-glutathione (DON-GSH)	OH	H	OH	OH	O	$C_{10}H_{17}N_3O_6S$
Fusarenon-X-3-*O*-glucoside (FUSX)	$C_6H_{12}O_6$	$OCOCH_3$	OH	OH	O	H

Figure 4.1   Basic chemical structure of a trichothecene and its respective modified mycotoxin forms.

standard of DON-3-Glc recently enabled the collection of occurrence data, albeit for this specific biologically modified, conjugated (masked) form only, whereas no quantitative information is available for other type A and B trichothecene glucosides.[32]

DON-3-Glc can be unambiguously identified using its reference standard; therefore, the mycotoxin has also been subject to analysis by HRMS (Figure 4.1). The aim of the study of Rubert *et al.* (2013) was to screen mycotoxins in European beers through an untargeted approach.[33] Using a hybrid LTQ orbitrap XL in positive electrospray mode (ESI$^+$), screening and identification was achieved. The structures of the mycotoxins were studied and the exact masses were calculated. DON-3-Glc, however, could not be identified in European beers.[33] Suman *et al.* (2012) reported the development of a LC/linear ion trap MS method capable of determining DON-3-Glc in different processed cereal-derived products.[34] The reliability of the method was finally demonstrated in bread, crackers, biscuits and mini-cake commodities, resulting in relatively low levels of DON-3-Glc, which were not higher than 30 μg kg^{-1}.[34]

Zachariasova *et al.* (2012) investigated the presence of DON-oligoglucosides in beer using a selective immunoaffinity-based preconcentration strategy, followed by ultra-high-performance liquid chromatography (UHPLC) coupled to HR-orbitrap MS.[35] The authors revealed that, in addition to the most common DON-3-Glc, oligoglucosylated DON with up to four bound hexose units was also present in beer and cereal-based products (Figure 4.1). The presence of DON di-, tri- and tetra-glucosides was demonstrated on the basis of accurate mass measurement and elemental composition calculation. The orbitrap was able to routinely achieve a mass resolving power as high as 65 000 FWHM at the *m/z* of DON-oligoglucosides, with calculated maximum mass errors not higher than 3 ppm.[35]

Besides DON, NIV frequently occurs in cereal-based food and feed. As these free mycotoxins have structural similarities, it was suggested that glucoside conjugates could be formed. Indeed, Nakagawa *et al.* (2012) revealed the presence of nivalenol-3-*O*-glucoside (NIV-3-Glc) in wheat artificially contaminated with *F. graminearum* (Figure 4.1).[36] Yoshinari *et al.* (2014) obtained similar results, and isolated NIV-3-Glc from NIV-contaminated wheat.[12] Compounds with an estimated molecular formula of $C_{15}H_{20}O_7$ and $C_{21}H_{30}O_{12}$ were screened with Q-TOF LC-MS and as a result of MS/MS fragmentation analysis, common fragment ions of NIV were found in the product ion spectra. The structure of NIV-3-Glc was confirmed with ^{1}H and ^{13}C nuclear magnetic resonance (NMR) spectroscopy. The identified percentage of NIV-3-Glc to NIV ranged from 12% to 27%.[12]

Furthermore, other studies revealed the presence of chemically modified, non-thermally formed DON-sulfonates. DON contamination can be reduced in animal feed by treatment with sodium bisulfite and sodium metabisulfite with the formation of DON-sulfonate resulting in a strongly reduced toxicity compared to DON. Three different DON-sulfonates, termed DON-sulfonate 1, 2 and 3, were identified and structurally elucidated by UHPLC coupled with HRMS/MS, as well as NMR spectroscopy (Figure 4.2). During MS

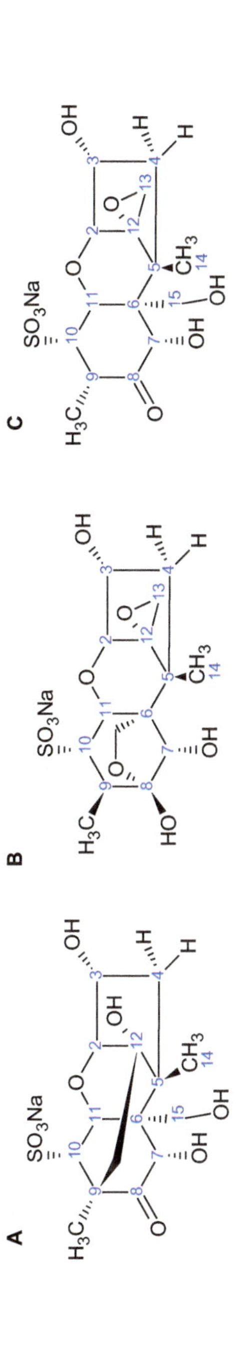

Figure 4.2 Chemical structures of the chemically modified, non-thermally formed DON-sulfonate 1 (A), DON-sulfonate 2 (B) and DON-sulfonate 3 (C).

experiments, ions were scanned in the range of *m/z* 100–1000, and data were stored in MS/MS experiments between *m/z* 50–550. UHPLC-HRMS measurements in ESI⁻ yielded the same accurate masses (378.0985 Da) and the molecular formula of $C_{15}H_{22}O_9S$ for all three DON-sulfonates.[15]

Recently, an untargeted screening strategy for the detection of biotransformation products of DON using stable isotopic labeling (SIL) and LC-HRMS was reported. The SIL-assisted approach was exemplified by the metabolization of DON *in planta*.[37] DON is a known virulence factor of *Fusarium*, and in turn, detoxification of DON seems to be an important component of *Fusarium* resistance in wheat. Plants can reduce the toxicity of DON by conjugation of the toxin to polar substances such as glucose.[37] The organism of interest (*i.e.* wheat) was treated with a mixture of the labeled and non-labeled precursor, DON. LC-HRMS measurements and data processing revealed a total of 57 corresponding peak pairs, which originated from ten DON biotransformation products. Besides the known DON and DON-3-Glc, which were confirmed by measurement of their reference standards, eight further DON biotransformation products were found by the untargeted screening approach. Based on a mass deviation of less than ± 5 ppm and MS/MS measurements, one of these products was annotated as the DON-glutathione (DON-GSH) conjugate (Figure 4.1), which was described for the first time in wheat. The data suggested that two DON-GSH-related conjugates, the processing products DON-S-cysteine and DON-S-cysteinyl-glycine and five unknown DON conjugates were formed *in planta*.[37]

Parallel to reports of conjugates of DON and NIV, Nakagawa *et al.* (2012) applied HRMS for the identification of another new *Fusarium* mycotoxin conjugate in wheat artificially contaminated with *F. graminearum*, namely fusarenon-X-3-*O*-glucoside (FUSX-3-Glc).[36] Identification was based on accurate mass measurements of characteristic ions and MS/MS fragmentation patterns. Quantification was not executable; however, 15% of the free form was estimated to be converted into the conjugated, masked form. The same authors were also able to detect neosolaniol-3-*O*-glucoside (NEO-3-Glc), diacetoxyscirpenol-3-*O*-glucoside (DAS-3-Glc)[14,36] and mono-acetoxyscirpenol-3-*O*-glucoside (MAS-3-Glc) in corn powder reference material (Figure 4.1).[14] Although the chemical structure was not clarified for these new modified mycotoxins, 3-*OH* glucosylation appeared to be most probable when considering the structures of FUSX, NEO, DAS and MAS, and the fragmentation profiles of these mycotoxins.

The mono-glucosides of T2 and HT2, produced by solid and liquid cultures of *Fusarium sporotrichioides*, were reported for the first time by Busman *et al.* (2011) using HPLC-MS/MS.[38] Along with the expected T2 and HT2, two additional compounds were detected, which had ions *m/z* 162 higher than those in the mass spectra of the free mycotoxins. Fragmentation behavior of these two compounds was similar to that of T2 and HT2. Based on LC-MS/MS behavior, it was proposed that the two compounds were T2-3-*O*-glucoside (T2-3-Glc) and HT2-3-*O*-glucoside (HT2-3-Glc) (Figure 4.1). Nakagawa *et al.* (2013) also reported the presence of these modified forms together with

HT2-3-*O*-diglucoside (HT2-3-diGlc) and T2-3-*O*-diglucoside (T2-3-diGlc) in corn powder reference material using orbitrap technology.[39] Samples were analyzed and the authors hypothesized that the mean contamination levels of T2-3-Glc and HT2-3-Glc were 24 μg kg^{-1} and 41 μg kg^{-1}, respectively, assuming that T2-3-Glc/T2 and HT2-3-Glc/HT2 was equal to the DON-3-Glc/DON ratio.

Using the same technology, T2-3-*O*-triglucoside (T2-3-triGlc) was reported, produced by a detached leaf *in vitro* model system, inoculated with the free mycotoxin T2.[40] In addition to this newly developed *in vitro* biosynthesis protocol, T2 glucosides were also isolated from *Fusarium*-infected plant material. McCormick *et al.* (2012) investigated an efficient microbial whole-cell catalytic method for the preparation of T2 glucosides with a potential large-scale availability of modified forms. Yeast species assigned to the *Trichomonascus* clade (*Saccharomycotina*, *Ascomycota*), classified in *Blastobotrys*, were tested for their ability to convert T2 to possible metabolization products.[41] The occurrence of 3-acetyl-T2 and T2-3-Glc and the removal of the isovaleryl group to form NEO were confirmed.

The occurrence of these two glucoside conjugates, T2-3-Glc and HT2-3-Glc, in naturally contaminated wheat and oats was clearly documented by Lattanzio *et al.* (2012b).[32] The use of an advanced orbitrap-based HRMS technology allowed for the obtaining of molecular structure details by measuring the accurate masses of the main characteristic fragments, with mass accuracy lower than 2.8 ppm (absolute value). T2-3-Glc, HT2-3-Glc and the monoglucoside HT2-4-*O*-glucoside (HT2-4-Glc) were identified and characterized. The analysis of their fragmentation patterns provided evidence for glucosylation at C-3 for T2 and at C-3 or C-4 for HT2. On the basis of the peak area ratio between glucoside conjugates and free T2 and HT2, the presence of glucosides was more likely in wheat than in oats samples. Conversion of 24% and 27% for T2 and HT2 glucosides, respectively, were observed.[32]

In contrast to orbitrap technology, TOF instruments were also used to confirm the postulated chemical formulae of mycotoxin conjugates. Veprikova *et al.* (2012) observed that T2-3-Glc and HT2-3-Glc occurred in 80% and 75% of the wheat samples with conversions of 16% to 17%, respectively.[42] The natural occurrence of HT2-3-diGlc was only observed in two samples using TOF-MS/MS. These quantitative differences were probably due to low recoveries of the analytes with the use of specific Mycosep$^{®}$ columns (Figure 4.1).

Plant conjugates of type C and D trichothecenes (*e.g.* verrucarin A-glucoside and roridine A, D and E-glucosides) were detected in *Baccharis coridifolia* by low-resolution LC-MS/MS. It remains unclear whether these conjugated, masked forms are less toxic storage forms or part of the mechanism of self-protection of the plant. The release of the toxic aglucons by glucosidases after exposure to herbivores might be required for animal toxicity.[43] Table 4.1 shows an overview of the literature concerning the screening and identification of modified trichothecenes using (HR)MS.

4.3.2 *Fusarium* Mycotoxins: Myco-estrogens

Zearalenone (ZEN) is the principal representative of the group of non-steroidal myco-estrogens. The structure consists of a phenolic resorcyclic acid lactone, chemically described as 3,4,5,6,9,10-hexahydro-14,16-dihydroxy-3-methyl-2-2-benzoxacyclotetradecin,1,7-dione. The main production of myco-estrogens is described for *F. graminearum*, *F. crookwellense*, *F. sporotrichioides* and *F. culmorum*, and consequently co-occurrence with DON and other trichothecenes is contingent.[44] The production, mainly in maize, as well as in other crops such as wheat, barley and oats, depends on environmental conditions and is favored by high humidity and low temperature, resulting in a wide geographical spread,[30,45,46] but also agricultural measures such as soil tillage and fungicide use can influence the presence of fusariotoxins.[47,48] The most abundant derivatives of ZEN are its functionalized, phase I metabolites, α-zearalenol (αZEL) and β-zearalenol (βZEL).

The capability of plants to metabolize ZEN was clearly observed by Berthiller *et al.* (2006).[49] Seventeen different biologically modified conjugates of ZEN were detected in ZEN-treated *Arabidopsis thaliana* by low-resolution LC-MS/MS. Known transformation products include zearalenone-14-glucoside (ZEN-14-Glc) (Figure 4.3), ZEN-14S, αZEL, βZEL, α-zearalenol-14-glucoside (αZEL-14-Glc) and β-zearalenol-14-glucoside (βZEL-14-Glc). Di-hexosides, hexose-pentosides and malonylglucosides of ZEN, αZEL and βZEL were also described.[49]

To date, only three research groups have investigated the occurrence of these biologically modified, conjugated myco-estrogens in food and feed using low-resolution LC-MS/MS, probably due to the unavailability of commercial reference standards.[50–52]

Schneweis *et al.* (2002) demonstrated that, out of a batch of 24 wheat samples, ZEN-14-Glc was found in 10 of the ZEN-positive samples (42%) at levels ranging from 17 μg kg^{-1} to 104 μg kg^{-1}.[50] The amounts of ZEN-14-Glc were correlated with those of ZEN, for which a correlation of 0.86 was found. Another research group analyzed 84 cereal-based products (wheat flour,

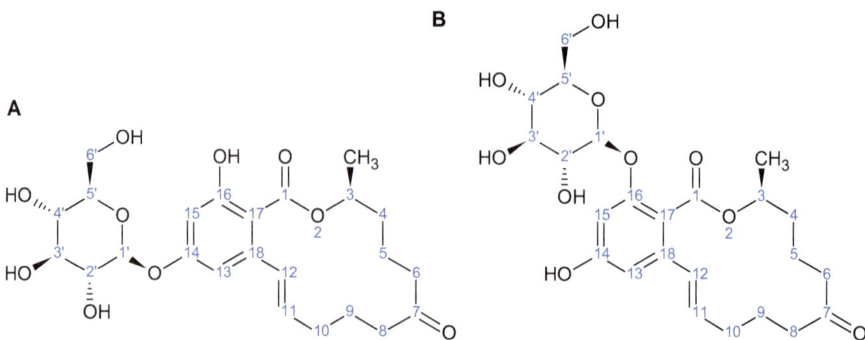

Figure 4.3 Chemical structures of the biologically modified, conjugated zearalenone-14-*O*-glucoside (A) and zearalenone-16-*O*-glucoside (B).

whole-meal wheat bread, maize meal, biscuits, wheat flakes, bran flakes, muesli, crackers, cereal snack bars and polenta), but none of the samples contained ZEN-14-Glc, αZEL, βZEL, αZEL-14-Glc or βZEL-14-Glc, notwithstanding ZEN-14-S being observed in low concentrations (6.1 µg kg^{-1}).[51]

De Boevre *et al.* (2012) analyzed 175 cereal-based foods, in which 30 samples contained ZEN, αZEL, βZEL, ZEN-14-Glc, αZEL-14-Glc, βZEL-14-Glc and ZEN-14-S.[52] The incidence of ZEN in food and feed matrices was 80%. αZEL and βZEL, respectively, occurred in 53% and 63% of the samples. ZEN-14-Glc (5%) was detected from trace levels up to 274 µg kg^{-1}. One maize sample proved the co-occurrence of ZEN-14-Glc (274 µg kg^{-1}), ZEN-14-S (51 µg kg^{-1}), βZEL-14-Glc (92 µg kg^{-1}) and a relatively low amount of ZEN (59 µg kg^{-1}), suggesting that approximately 90% of the original ZEN had been metabolized.[52]

Recently, Kovalsky-Paris *et al.* (2014) confirmed the presence of a new myco-estrogen, zearalenone-16-*O*-glucoside (ZEN-16-Glc) (Figure 4.3).[53] LC-HRMS/MS spectra were recorded over a range of *m/z* 50–1000 after positive and negative electrospray ionization. A substance of $C_{24}H_{32}O_{10}$ (*m/z* 481.2064) was isolated and a conjugate of ZEN was hypothesized. MS/MS spectra showed the cleavage of a glucose moiety; however, further fragments from ZEN could not be used because there was no chromatographic separation between ZEN-14-Glc and ZEN-16-Glc. Therefore, ^{1}H and ^{13}C NMR were used to confirm and elucidate the structure of ZEN-16-Glc.[53]

4.4 Untargeted Analysis of Modified Mycotoxins in Human Biological Fluids

Over the years, several studies have reported the use of triple-quadrupole MS for the detection of mycotoxin biomarkers in human biological fluids. Meanwhile, analysis has been limited to only the known free mycotoxins. This has been attributed to the lack of reference standards and/or the non-availability of advanced instrumentation such as HRMS for the identification of new metabolites. While the free mycotoxins such as DON, ochratoxin A (OTA), ZEN, fumonisin B$_1$ (FB$_1$) and aflatoxin B$_1$ (AFB$_1$) have been extensively studied by low-resolution MS, several other phase I or phase II metabolites of these analytes have often escaped routine surveillance. The application of HRMS in the mycotoxin biomarker research field is scarce to non-existent. The following summary and discussion will provide the reader with a general overview of some of the most important metabolites belonging to this specific class of mycotoxins in human biological fluids by the use of MS/MS and HRMS.

4.4.1 *Fusarium* Mycotoxins

DON is almost exclusively excreted as a biologically modified, conjugated glucuronide accounting for about 91% of total DON.[54] DON and its detoxification phase II metabolite DON-3-GlcA have often been reported in the urine of exposed humans. Meanwhile, studies carried out in the United Kingdom showed a strong correlation between these urinary metabolites (*i.e.* the sum of the free DON and DON-3-GlcA) and cereal intake of the study population.[54] As a consequence, DON-3-GlcA was suggested as a biomarker for the assessment of human dietary exposure to this toxin.[54–56] In rat urine, the predominating glucuronide was suggested to be DON-3-GlcA.[57] Recently, Warth *et al.* described a second glucuronide dominating in human urine, tentatively assigned as being DON-15-glucuronide (DON-15-GlcA).[58] In a microsomal study using rat and pig liver microsomes, no hepatic metabolism of DON was observed.[59] By contrast, a study performed with human, rat and minipig microsomes led to the qualitative detection of up to three glucuronides. The third glucuronide was postulated and later confirmed to be DON-7-glucuronide (DON-7-GlcA) by NMR. Since DON possesses one primary (C-15) and two secondary (C-3 and C-7) hydroxyl groups, it was obvious to assume that the third glucuronide was the 7-*O*-glucuronide of DON. This hypothesis was supported by the LC-MS and HRMS characteristics of the compound.[60] Recently, a fourth glucuronide with a possible hydroxylation point at position C-8 was described and confirmed by MS and NMR data.[61]

T2, which belongs to the family of type A trichothecenes, is rapidly metabolized by esterases, resulting in several metabolites being detected *in vivo* and *in vitro*. The spectrum and the ratios of T2 metabolites in animals strongly depend on the investigated species.[62] The main biotransformation pathway is deacetylation of the C-4 acetyl group to HT2. Other metabolites detected after incubation of T2 with Chinese hamster ovary cells and African green monkey kidney cells included traces of T2 triol and T2 tetraol.[63] A possible metabolism of T2 to NEO by carboxylesterase activity in human blood cells has been reported.[64] In these cells, both HT2 and NEO were produced in equal amounts. Other metabolites were T2 tetraol and several hydroxylated derivatives such as 3-OH-T2 and 3-OH-HT2. In addition to these metabolites, several other postulated conjugates have been reported.[65] Thus, with regard to tissues, body fluids and feces, efforts to identify and characterize the time course of the principal metabolites of T2 may be of greater diagnostic value than analysis of the free compound.

The first reported mammalian biologically modified mycotoxins for ZEN were the stereo-isomers αZEL, βZEL, α-zearalanol (αZAL) and β-zearalanol (βZAL), assumed to be catalyzed by 3α- and 3β-hydroxyl-steroid-dehydrogenase, and the conjugation of these stereo-isomers with glucuronic acid.[66] Furthermore, functionalized conjugation of ZEN to proteins such as serum albumin has not been investigated. 13-Hydroxy-ZEN and 15-hydroxy-ZEN, arising through aromatic hydroxylation of ZEN, have been reported using human hepatic microsomes.[67] Furthermore, a monohydroxylated ZEN metabolite, tentatively identified as 8-hydroxy-ZEN, has recently been detected in liver microsomal preparations from various animal species, and also in the liver and urine of rats dosed with ZEN.[68] All of the human biomonitoring programs have been limited to the detection of only ZEN and the stereo-isomers (αZEL, αZAL, βZEL and βZAL).

To evaluate the exposure to FB_1 in experimental animal studies, the ratio of sphinganine to sphingosine in urine and serum was reported as a functional biomarker;[69] however, this failed to yield the expected outcome with human studies.[70–72] FB_1 is reported not to undergo any major metabolism, as incubation with primary rat hepatocyte cultures and subcellular enzyme fractions failed to produce detectable metabolites.[72] Furthermore, FB_1 was recovered unaltered in the urine, feces and bile of dosed animals. Hydrolysis of the two tricarballylic acid ester groups of FB_1 has been reported to occur in the gut of vervet monkeys,[66,73] and could possibly serve as an alternative biomarker. Hydrolyzed FB_1 has not been reported in human cell culture studies and has not been detected in human urine samples.

4.5 Conclusion

In the first section of this chapter, the many possibilities of HRMS as a tool in (modified) mycotoxin analysis were pointed out. The state of the art of analytical chemistry using HRMS permits accurate mass measurements in many fields of (modified) mycotoxin analysis; however, it should be stated that confirmation and structure elucidation of new unknown molecules with ^1H and ^{13}C NMR remains imperative. Based on the existing literature on biologically and chemically modified mycotoxins, it can be concluded that HRMS is becoming a widespread tool for the identification of these mycotoxins.

UHPLC is nowadays the separation tool prior to MS detection, and offers positive impacts on system efficiency and analysis time. The implementation of UHPLC-HRMS in routine analysis is an interesting growing field and offers great potential. To date, focus has been towards multi-mycotoxin methods with the inclusion of more than 20 mycotoxins,[74–76] however, by using HRMS, the number of analytes can be extended to 400 xenobiotics and more.[77]

The implementation of HRMS in the research field of mycotoxin biomarkers in human biological fluids is scarce to non-existent. However, based on the possibilities offered by HRMS, the technique could offer great potential in the future.

In conclusion, HRMS is a very useful tool with a high potential for efficient automated screening in the field of food and feed safety and the untargeted analysis of new modified mycotoxins.

References

1. M. Rychlik, H. U. Humpf, D. Marko, S. Danicke, A. Mally, F. Berthiller, H. Klaffke and N. Lorenz, Proposal of a comprehensive definition of modified and other forms of mycotoxins including "masked" mycotoxins, *Mycotoxin Res*, 2014, **30**, 197–205.

2. M. Zachariasova, T. Cajka, M. Godula, A. Malachova, Z. Veprikova and J. Hajslova, Analysis of multiple mycotoxins in beer employing (ultra)-high-resolution mass spectrometry, *Rapid Commun. Mass Spectrom.*, 2010, **24**, 3357–3367.

3. J. Rubert, K. J. James, J. Manes and C. Soler, Applicability of hybrid linear ion trap-high resolution mass spectrometry and quadrupole-linear ion trap-mass spectrometry for mycotoxin analysis in baby food, *J. Chromatogr. A*, 2012, **1223**, 84–92.

4. K. K. Murray, R. K. Boyd, M. N. Eberlin, G. J. Langley, L. Li and Y. Naito, Definitions of terms relating to mass spectrometry (IUPAC Recommendations 2013), *Pure Appl. Chem.*, 2013, **85**, 1515–1609.

5. J. Jennessen, K. F. Nielsen, J. Houbraken, E. K. Lyhne, J. Schnurer, J. C. Frisvad and R. A. Samson, Secondary metabolite and mycotoxin production by the Rhizopus microsporus group, *J. Agric. Food Chem.*, 2005, **53**, 1833–1840.

6. V. M. T. Lattanzio, A. Visconti, M. Haidukowski and M. Pascale, Identification and characterization of new *Fusarium* masked mycotoxins, T-2 and HT-2 glycosyl derivatives, in naturally contaminated wheat and oats by liquid chromatography high-resolution mass spectrometry, *J. Mass Spectrom.*, 2012, **47**, 466–475.

7. V. M. Scussel, M. Rokka, A. Rizzo, M. Jestoi and K. Peltonen, Characterization of DON and DON-3-beta-D-glucopyranoside through accurate mass measurement by quadrupole-time-of-flight mass spectrometry, *Int. J. Environ. Anal. Chem.*, 2013, **93**, 61–74.

8. Z. Veprikova, M. Vaclavikova, O. Lacina, Z. Dzuman, M. Zachariasova and J. Hajslova, Occurrence of mono- and di-glycosylated conjugates of T-2 and HT-2 toxins in naturally contaminated cereals, *World Mycotoxin J.*, 2012, **5**, 231–240.

9. B. Kluger, C. Bueschl, M. Lemmens, F. Berthiller, G. Haubl, G. Jaunecker, G. Adam, R. Krska and R. Schuhmacher, Stable isotopic labelling-assisted untargeted metabolic profiling reveals novel conjugates of the mycotoxin deoxynivalenol in wheat, *Anal. Bioanal. Chem.*, 2013, **405**, 5031–5036.

10. H. Nakagawa, K. Ohmichi, S. Sakamoto, Y. Sago, M. Kushiro, H. Nagashima, M. Yoshida and T. Nakajima, Detection of a new Fusarium masked mycotoxin in wheat grain by high-resolution LC-Orbitrap MS, *Food Addit. Contam., Part A*, 2011, **28**, 1447–1456.

11. H. Nakagawa, S. Sakamoto, Y. Sago and H. Nagashima, Detection of Type A Trichothecene Di-Glucosides Produced in Corn by High-Resolution Liquid Chromatography-Orbitrap Mass Spectrometry, *Toxins*, 2013, **5**, 590–604.

12. T. Yoshinari, S. Sakuda, K. Furihata, H. Furusawa, T. Ohnishi, Y. Sugita-Konish, N. Ishizaki and J. Terajima, Structural Determination of a Nivalenol Glucoside and Development of an Analytical Method for the Simultaneous Determination of Nivalenol and Deoxynivalenol, and Their Glucosides, in Wheat, *J. Agric. Food Chem.*, 2014, **62**, 1174–1180.

13. M. Zachariasova, M. Vaclavikova, O. Lacina, L. Vaclavik and J. Hajslova, Deoxynivalenol Oligoglycosides: New Masked Fusarium Toxins Occurring in Malt, Beer, and Breadstuff, *J. Agric. Food Chem.*, 2012, **60**, 9280–9291.
14. H. Nakagawa, S. Sakamoto, Y. Sago, M. Kushiro and H. Nagashima, Detection of masked mycotoxins derived from type A trichothecenes in corn by high-resolution LC-Orbitrap mass spectrometer, *Food Addit. Contamin., A*, 2013, **30**, 1407–1414.
15. H. E. Schwartz, C. Hametner, V. Slavik, O. Greitbauer, G. Bichl, E. Kunz-Vekiru, D. Schatzmayr and F. Berthiller, Characterization of Three Deoxynivalenol Sulfonates Formed by Reaction of Deoxynivalenol with Sulfur Reagents, *J. Agric. Food Chem.*, 2013, **61**, 8941–8948.
16. R. Gunnaiah, A. C. Kushalappa, R. Duggavathi, S. Fox and D. J. Somers, ntegrated Metabolo-Proteomic Approach to Decipher the Mechanisms by Which Wheat QTL (Fhb1) Contributes to Resistance against Fusarium graminearum, *Plos One*, 2012, 7, e40695.
17. S. V. Malysheva, N. Arroyo-Manzanares, J. W. Cary, K. C. Ehrlich, J. Vanden Bussche, L. Vanhaecke, D. Bhatnagar, J. D. Di Mavungu and S. De Saeger, Identification of novel metabolites from Aspergillus flavus by high resolution and multiple stage mass spectrometry, *Food Addit. Contamin., A*, 2014, **31**, 111–120.
18. L. Monaci, E. De Angelis and A. Visconti, Determination of deoxynivalenol, T-2 and HT-2 toxins in a bread model food by liquid chromatography-high resolution-Orbitrap-mass spectrometry equipped with a high-energy collision dissociation cell, *J. Chromatogr. A*, 2011, **1218**, 8646–8654.
19. C. Van Poucke, C. Detavernier and C. Van Peteghem in *Food Analysis by HPLC*, ed. L. M. L. Nollet, Taylor & Francis, 2012.
20. E. De Dominicis, I. Commissati and M. Suman, Targeted screening of pesticides, veterinary drugs and mycotoxins in bakery ingredients and food commodities by liquid chromatography-high-resolution single-stage Orbitrap mass spectrometry, *J. Mass Spectrom.*, 2012, **47**, 1232–1241.
21. S. M. Lehner, N. K. N. Neumann, M. Sulyok, M. Lemmens, R. Krska and R. Schuhmacher, Evaluation of LC-high-resolution FT-Orbitrap MS for the quantification of selected mycotoxins and the simultaneous screening of fungal metabolites in food, *Food Addit. Contamin., A*, 2011, **28**, 1457–1468.
22. J. Rubert, Z. Dzuman, M. Vaclavikova, M. Zachariasova, C. Soler and J. Hajslova, Analysis of mycotoxins in barley using ultra high liquid chromatography high resolution mass spectrometry: Comparison of efficiency and efficacy of different extraction procedures, *Talanta*, 2012, **99**, 712–719.
23. H. G. J. Mol, P. Plaza-Bolanos, P. Zomer, T. C. de Rijk, A. A. M. Stolker and P. P. J. Mulder, Toward a Generic Extraction Method for Simultaneous Determination of Pesticides, Mycotoxins, Plant Toxins, and

Veterinary Drugs in Feed and Food Matrixes, *Anal. Chem.*, 2008, **80**, 9450–9459.

24. S. De Baere, A. Osselaere, M. Devreese, L. Vanhaecke, P. De Backer and S. Croubels, Development of a liquid-chromatography tandem mass spectrometry and ultra-high-performance liquid chromatography high-resolution mass spectrometry method for the quantitative determination of zearalenone and its major metabolites in chicken and pig plasma, *Anal. Chim. Acta*, 2012, **756**, 37–48.

25. J. Rubert, J. Manes, K. J. James and C. Soler, Application of hybrid linear ion trap-high resolution mass spectrometry to the analysis of mycotoxins in beer, *Food Addit. Contamin., A*, 2011, **28**, 1438–1446.

26. F. Berthiller, C. Crews, C. Dall'Asta, S. D. Saeger, G. Haesaert, P. Karlovsky, I. P. Oswald, W. Seefelder, G. Speijers and J. Stroka, Masked mycotoxins: a review, *Mol. Nutr. Food Res.*, 2013, 57(1), 165–186.

27. J. F. Grove in *Progress in the Chemistry of Organic Natural Products*, ed. W. Herz, H. Falk and G. W. Kirby, Springer, Vienna, 2007, pp. 63–130.

28. Y. Ueno, The toxicology of mycotoxins, *Crit. Rev. Toxicol.*, 1985, **14**, 99–132.

29. U. Thrane, A. Adler, P. E. Clasen, F. Galvano, W. Langseth, A. Logrieco, K. F. Nielsen and A. Ritieni, Diversity in metabolite production by Fusarium langsethiae, Fusarium poae, and Fusarium sporotrichioides, *Int. J. Food Microbiol.*, 2004, **95**, 257–266.

30. E. Richard, N. Heutte, L. Sage, D. Pottier, V. Bouchart, P. Lebailly and D. Garon, Toxigenic fungi and mycotoxins in mature corn silage, *Food Chem. Toxicol.*, 2007, **45**, 2420–2425.

31. T. Brasel, M. Martin, C. Carriker, S. Wilson and D. Straus, Macrocyclic Trichothecene Mycotoxins in the Indoor Environment, *Appl. Environ. Microbiol.*, 2005, **71**, 7376–7388.

32. V. M. T. Lattanzio, A. Visconti, M. Haidukowski and M. Pascale, Identification and characterization of new *Fusarium* masked mycotoxins, T-2 and HT-2 glycosyl derivatives, in naturally contaminated wheat and oats by liquid chromatography high-resolution mass spectrometry, *J. Mass Spectrom.*, 2012, **47**, 466–475.

33. J. Rubert, C. Soler, R. Marin, K. J. James and J. Manes, Mass spectrometry strategies for mycotoxins analysis in European beers, *Food Contam.*, 2013, **30**, 122–128.

34. M. Suman, E. Bergamini, D. Catellani and A. Manzitti, Development and validation of a liquid chromatography/linear ion trap mass spectrometry method for the quantitative determination of deoxynivalenol-3-glucoside in processed cereal-derived products, *Food Chem.*, 2013, **136**, 1568–1576.

35. M. Zachariasova, M. Vaclavikova, O. Lacina, L. Vaclavik and J. Hajslova, Deoxynivalenol Oligoglycosides: New Masked Fusarium Toxins Occurring in Malt, Beer, and Breadstuff, *J. Agric. Food Chem.*, 2012, **60**, 9280–9291.

36. H. Nakagawa, S. Sakamoto, Y. Sago, M. Kushiro and H. Nagashima, The use of LC-Orbitrap MS for the detection of *Fusarium* masked mycotoxins: the case of type A trichothecenes, *World Mycotoxin J.*, 2012, **5**, 271–280.

37. B. Kluger, C. Bueschl, M. Lemmens, F. Berthiller, G. Haubl, G. Jaunecker, G. Adam, R. Krska and R. Schuhmacher, Stable isotopic labelling-assisted untargeted metabolic profiling reveals novel conjugates of the mycotoxin deoxynivalenol in wheat, *Anal. Bioanal. Chem.*, 2013, **405**, 5031–5036.

38. M. Busman, S. Poling and C. Maragos, Observation of T-2 toxin and HT-2 toxin glucosides from *Fusarium sporotrichioides* by liquid chromatography coupled to tandem mass spectrometry (LC-MS/MS), *Toxins*, 2011, **3**, 1554–1568.

39. H. Nakagawa, S. Sakamoto, Y. Sago and H. Nagashima, Detection of Type A Trichothecene Di-Glucosides Produced in Corn by High-Resolution Liquid Chromatography-Orbitrap Mass Spectrometry, *Toxins*, 2013, **5**, 590–604.

40. M. De Boevre, A. Vanheule, K. Audenaert, B. Bekaert, J. Diana Di Mavungu, S. Werbrouck, G. Haesaert and S. De Saeger, Detached leaf *in vitro* model for masked mycotoxin biosynthesis and subsequent analysis of unknown conjugates, *World Mycotoxin J.*, 2014, **7**, 305–312.

41. S. P. McCormick, N. P. J. Price and C. P. Kurtzman, Glucosylation and other biotransformations of T-2 toxin by yeasts of the *Trichomonascus clade*, *Appl. Environ. Microbiol.*, 2012, **78**, 8694–8702.

42. Z. Veprikova, M. Vaclavikova, O. Lacina, Z. Dzuman, M. Zachariasova and J. Hajslova, Occurrence of mono- and di-glycosylated conjugates of T-2 and HT-2 toxins in naturally contaminated cereals, *World Mycotoxin J.*, 2012, **5**, 231–240.

43. F. Berthiller, R. Schuhmacher, G. Adam and R. Krska, Formation, determination and significance of masked and other conjugated mycotoxins, *Anal. Bioanal. Chem.*, 2009, **395**, 1243–1252.

44. A. Pittet, Natural occurrence of mycotoxins in foods and feeds-an updated review, *Rev. Méd. Vét.*, 1998, **149**, 479–492.

45. M. Peraica, B. Radic, A. Lucic and M. Pavlovic, Toxic effects of mycotoxins in humans, *Bull. W. H. O.*, 1999, **77**, 754–766.

46. S. Yazar and G. Z. Omurtag, Fumonisins, Trichothecenes and Zearalenone in Cereals, *Int. J. Mol. Sci.*, 2008, **9**, 2062–2090.

47. K. Audenaert, E. Callewaert, S. De Saeger, M. Höfte and G. Haesaert, Hydrogen peroxide induced by application of triazole fungicides triggers deoxynivalenol (DON) production by *Fusarium graminearum*, *BMC Microbiol.*, 2010, **10**, 1–14.

48. S. Landschoot, W. Waegeman, K. Audenaert, J. Vandepitte, J. Baetens, G. Haesaert and B. De Baets, An empirical analysis of explanatory variables affecting *Fusarium* infection and deoxynivalenol production in wheat (2010), *J. Plant Pathol.*, 2012, **94**, 135–147.

49. F. Berthiller, U. Werner, M. Sulyok, R. Krska, M. T. Hauser and R. Schuhmacher, Liquid chromatography coupled to tandem mass spectrometry (LC-MS/MS) determination of phase II metabolites of the mycotoxin zearalenone in the model plant *Arabidopsis thaliana*, *Food Addit. Contamin.*, 2006, **23**, 1194–1200.

50. I. Schneweis, K. Meyer, G. Engelhardt and J. Bauer, Occurrence of zearalenone-4-beta-D-glucopyranoside in wheat, *J. Agric. Food Chem.*, 2002, **50**, 1736–1738.
51. O. Vendl, C. Crews, S. MacDonald, R. Krska and F. Berthiller, Occurrence of free and conjugated Fusarium mycotoxins in cereal-based food, *Food Addit. Contamin., A*, 2010, **27**, 1148–1152.
52. M. De Boevre, J. Diana Di Mavungu, S. Landschoot, K. Audenaert, M. Eeckhout, P. Maene, G. Haesaert and S. De Saeger, Natural occurrence of mycotoxins and their masked forms in food and feed products, *World Mycotoxin J.*, 2012, **5**, 207–219.
53. M. P. Kovalsky-Paris, W. Schweiger, C. Hametner, R. Stuckler, G. J. Muehlbauer, E. Varga, R. Krska, F. Berthiller and G. Adam, Zearalenone-16-O-glucoside: A New Masked Mycotoxin, *J. Agric. Food Chem.*, 2014, **62**, 1181–1189.
54. P. C. Turner, V. J. Burley, J. A. Rothwell, K. L. M. White, J. E. Cade and C. P. Wild, Deoxynivalenol: Rationale for development and application of a urinary biomarker, *Food Addit. Contamin., A*, 2008, **25**, 864–871.
55. B. Warth, M. Sulyok, F. Berthiller, R. Schuhmacher, P. Fruhmann, C. Hametner, G. Adam, J. Frohlich and R. Krska, Direct quantification of deoxynivalenol glucuronide in human urine as biomarker of exposure to the Fusarium mycotoxin deoxynivalenol, *Anal. Bioanal. Chem.*, 2011, **401**, 195–200.
56. B. Warth, M. Sulyok, P. Fruhmann, F. Berthiller, R. Schuhmacher, C. Hametner, G. Adam, J. Frohlich and R. Krska, Assessment of human deoxynivalenol exposure using an LC-MS/MS based biomarker method, *Toxicol. Lett.*, 2012, **211**, 85–90.
57. V. M. T. Lattanzio, M. Solfrizzo, A. De Girolamo, S. N. Chulze, A. M. Torres and A. Visconti, LC-MS/MS characterization of the urinary excretion profile of the mycotoxin deoxynivalenol in human and rat, *J. Chromatogr. B*, 2011, **879**, 707–715.
58. B. Warth, M. Sulyok, P. Fruhmann, H. Mikula, F. Berthiller, R. Schuhmacher, C. Hametner, W. A. Abia, G. Adam, J. Frohlich and R. Krska, Development and validation of a rapid multi-biomarker liquid chromatography/tandem mass spectrometry method to assess human exposure to mycotoxins, *Rap. Commun. Mass Spectrom.*, 2012, **26**, 1533–1540.
59. L. M. Cote, W. Buck and E. Jeffery, Lack of Hepatic-Microsomal Metabolism of Deoxynivalenol and Its Metabolite, Dom-1, *Food Chem. Toxicol.*, 1987, **25**, 291–295.
60. R. Maul, C. Müller, S. Rieβ, M. Koch, F. J. Methner and I. Nehls, Germination induces the glucosylation of the Fusarium mycotoxin deoxynivalenol in various grains, *Food Chem.*, 2012, **131**, 274–279.
61. S. Uhlig, L. Ivanova and C. K. Faeste, Characterization of the 3-, 8-, and 15-Glucuronides of Deoxynivalenol, *J. Agric. Food Chem.*, 2013, **61**, 2006–2012.
62. B. Yagen and M. Bialer, Metabolism and Pharmacokinetics of T-2 Toxin and Related Trichothecenes, *Drug Metabol. Rev.*, 1993, **25**, 281–323.

63. L. R. Trusal, Metabolism of T-2 Mycotoxin by Cultured-Cells, *Toxicon*, 1986, **24**, 597–603.
64. H. Johnsen, E. Odden, B. A. Johnsen and F. Fonnum, Metabolism of T-2 Toxin by Blood-Cell Carboxylesterases, *Biochem. Pharmacol.*, 1988, **37**, 3193–3197.
65. C. C. Abnet, C. B. Borkowf, Y. L. Qiao, P. S. Albert, E. Wang, A. H. Merrill, S. D. Mark, Z. W. Dong, P. R. Taylor and S. M. Dawsey, Sphingolipids as biomarkers of fumonisin exposure and risk of esophageal squamous cell carcinoma in China, *Cancer, Causes Control*, 2001, **12**, 821–828.
66. A. Zinedine, J. M. Soriano, J. C. Molto and J. Manes, Review on the toxicity, occurrence, metabolism, detoxification, regulations and intake of zearalenone: An oestrogenic mycotoxin, *Food Chem. Toxicol.*, 2007, **45**, 1–18.
67. E. Pfeiffer, A. Hildebrand, H. Mikula and M. Metzler, Glucuronidation of zearalenone, zeranol and four metabolites in vitro: Formation of glucuronides by various microsomes and human UDP-glucuronosyltransferase isoforms, *Mol. Nutr. Food Res.*, 2010, **54**, 1468–1476.
68. F. Bravin, R. C. Duca, P. Balaguer and M. Delaforge, In Vitro Cytochrome P450 Formation of a Mono-Hydroxylated Metabolite of Zearalenone Exhibiting Estrogenic Activities: Possible Occurrence of This Metabolite in Vivo, *Int. J. Mol. Sci.*, 2009, **10**, 1824–1837.
69. L. van der Westhuizen, N. L. Brown, W. F. O. Marasas, S. Swanevelder and G. S. Shephard, Sphinganine/sphingosine ratio in plasma and urine as a possible biomarker for fumonisin exposure in humans in rural areas of Africa, *Food Chem. Toxicol.*, 1999, **37**, 1153–1158.
70. M. Solfrizzo, S. N. Chulze, C. Mallmann, A. Visconti, A. De Girolamo, F. Rojo and A. Torres, Comparison of urinary sphingolipids in human populations with high and low maize consumption as a possible biomarker of fumonisin dietary exposure, *Food Addit. Contamin., A*, 2004, **21**, 1090–1095.
71. L. J. G. Silva, C. M. Lino and A. Pena, Sphinganine-sphingosine ratio in urine from two Portuguese populations as biomarker to fumonisins exposure, *Toxicon*, 2009, **54**, 390–398.
72. M. E. Cawood, W. C. A. Gelderblom, J. F. Alberts and S. D. Snyman, nteraction of C-14-Labeled Fumonisin-B Mycotoxins with Primary Rat Hepatocyte Cultures, *Food Chem. Toxicol.*, 1994, **32**, 627–632.
73. G. S. Shephard, P. G. Thiel and E. W. Sydenham, Liquid-Chromatographic Determination of the Mycotoxin Fumonisin B-2 in Physiological Samples, *J. Chromatogr. A*, 1995, **692**, 39–43.
74. S. Monbaliu, C. Van Poucke, C. Van Peteghem, K. Van Poucke, K. Heungens and S. De Saeger, Development of a multi-mycotoxin liquid chromatography/tandem mass spectrometry method for sweet pepper analysis, *Rapid Commun. Mass Spectrom.*, 2009, **23**, 3–11.
75. M. Sulyok, F. Berthiller, R. Krska and R. Schuhmacher, Development and validation of a liquid chromatography/tandem mass spectrometric method for the determination of 39 mycotoxins in wheat and maize, *Rapid Commun. Mass Spectrom.*, 2006, **20**, 2649–2659.

76. E. Varga, T. Glauner, F. Berthiller, R. Krska, R. Schuhmacher and M. Sulyok, Development and validation of a (semi-)quantitative UHPLC-MS/MS method for the determination of 191 mycotoxins and other fungal metabolites in almonds, hazelnuts, peanuts and pistachios, *Anal. Bioanal. Chem.*, 2013, **405**, 5087–5104.
77. M. Kellman, H. Muenster, P. Zomer and H. Mol, Development and validation of a (semi-)quantitative UHPLC-MS/MS method for the determination of 191 mycotoxins and other fungal metabolites in almonds, hazelnuts, peanuts and pistachios, *J. Am. Soc. Mass Spectrom.*, 2009, **20**, 1464–1476.

Transformation of Mycotoxins upon Food Processing: Masking, Binding and Degradation Phenomena

MICHELE SUMAN* AND SILVIA GENEROTTI

Advanced Laboratory Research, via Mantova 166, 43122 Parma, Italy
*Email: michele.suman@barilla.com

Sorting, cleaning, milling and thermal processes can potentially alter the concentration of mycotoxins because of mechanical or thermal energies that in fact determine transformation and/or degradation. Data about the impact of food processing on masked/modified mycotoxins are, however, very limited. This chapter will focus on the potential transformation of (masked/modified) mycotoxins during food processing, as well as on possible amendment strategies by food processing.

5.1 Pre-milling

Harvested kernels are converted into flour for human consumption. The primary processing consists of selection and pre-milling steps. No significant thermal breakdown of mycotoxins would be expected at this stage, but moulds and mycotoxins are often concentrated in dust and broken grains that are more susceptible to fungal infection and toxin contamination. The removal of this material can result in a lowering of the contamination level,

Issues in Toxicology No. 24
Masked Mycotoxins in Food: Formation, Occurrence and Toxicological Relevance
Edited by Chiara Dall'Asta and Franz Berthiller

Published by the Royal Society of Chemistry, www.rsc.org

and the lowering of the contamination level may depend on the quality of the grain on receipt or how the mycotoxin is distributed within the individual grains. Sorting, cleaning, dehulling and debranning prior to milling may reduce the mycotoxin contamination in the grains by removing the kernels with extensive mould growth, broken kernels, fine materials and dust.[1,2] The results indicate that the effects of the pre-milling processes and the efficiency of the mycotoxin removal are extremely variable.

In various studies, the cleaning appears to be relatively ineffective for the mycotoxin deoxynivalenol (DON).[3,4] However, several authors reported that the concentration of DON in the cleaned cereals compared to the concentration of DON in cereals that have not been cleaned ranged from 7% to 63% for DON, from 7% to almost 100% for nivalenol (NIV) and from 7% to 40% for zearalenone (ZEN),[5,6] with a high mycotoxin concentration in the screening and dense fractions as a result of *Fusarium graminearum* fungus infection from the outside of the kernel.[7–12] For T-2 and HT-2 toxins, cleaning of the raw material did not lead to significant reductions of the mycotoxin levels, whereas dehulling led to a reduction of over 90% and T-2 toxin was found exclusively at low levels in the kernels, indicating that HT-2 toxin is more strictly confined to oat hulls.[13] The cleaning process is also the first way of removing fumonisins (FBs) concentrated on the pericarp of the kernel and in the broken and damaged grains: the step of cleaning maize can reduce FB residues to less than half that of uncleaned maize.[14] Sydenham described a FB reduction of 26–69% in corn after cleaning.[15] FB reduction was also confirmed by Scudamore *et al.*, studying the effects of cleaning on the levels of aflatoxins (AFs), ZEN and FBs in maize in a large commercial mill.[16] AF and FB concentrations were reduced by approximately 30–40%, but little reduction was observed in the levels of ZEN. Several factors, such as the initial condition of the grains, the type and extent of the contamination and the type of cleaning process, may be involved in this response.

The debranning process was widely investigated by several authors who observed a reduction of DON in debranned wheat ranging from 15 to 78%,[17–19] depending on the length of the pearling process and the percentage of grain tissue removal, and not on the starting contamination level in the raw material.[20] Moreover, Aureli and D'Egidio reported that debranning before milling was more efficient than the classical milling process for reducing mycotoxin content in by-products.[17] Table 5.1 summarises the effects of pre-milling on (masked) mycotoxins.

5.2 Milling

Whole grains are milled to produce flour and other fractions, then these fractions are the raw materials used to prepare the final products such as: (i) bread, biscuits and cakes by baking; (ii) breakfast cereals and snack products by extrusion; and (iii) porridge, polenta and similar products by boiling or steaming. As in cleaning and debranning, the milling process has no step

Table 5.1   Effects of pre-milling on (masked) mycotoxins.

Mycotoxin(s); matrix/commodity	Ref.	Food processing; behaviour/effects
All	1	Review
DON	2	Review
DON (cleaning)	3	No modifications
DON (cleaning)	4	No modifications
DON/NIV/ZEN (cleaning)	5	DON: reduction 7–63%; NIV: reduction 7–100%; ZEN: reduction 7–40%
DON/NIV/ZEN (cleaning)	6	DON: reduction 7–63%; NIV: reduction 7–100%; ZEN: reduction 7–40%
DON (cleaning)	7	Reduction in semolina
DON (cleaning)	8	Concentration in dockage
DON (cleaning)	9	Concentration in screening
DON (cleaning)	10	Increase in bran
DON (cleaning)	11	Increase in the dense fraction
DON (cleaning)	12	Concentration in bran
T-2/HT-2	13	Cleaning: no modifications Dehulling: reduction 90%
FBs (cleaning)	14	Reduction up to 50%
FBs (cleaning)	15	Reduction 26–69%
AFs/ZEN/FBs	16	AFs/FBs: reduction 20%/30% ZEN: no modification
DON (debranning)	17	Reduction 15–78%
DON (debranning)	18	Reduction 15–78%
DON (debranning)	19	Reduction 15–78%
DON (debranning)	20	Reduction 15–78%

that destroys mycotoxins; however, mycotoxin contamination may be redistributed and concentrated in certain milling fractions, even if mycotoxins can never be completely removed by processing.[21] Most mycotoxins tend to concentrate in the bran fractions or outer layers of the grains so that other parts of the cereal structure that produce fractions, such as white flour or maize grits, are usually contaminated with lower concentrations of mycotoxin than the fractions or outer layers that are present in the original whole grain. These cereal milling fractions are less likely to be used for food production but are used mainly as animal feeds. At the same time, these cereal milling fractions represent a novel category of promising ingredients for human nutrition and health, due to other interesting functional properties.

Extensive data are obtained by laboratory-scale test mills: even taking into account the fact that the approach in each experiment was different and the variability of the data was very high, several studies reported similar trends in mycotoxin distribution in the different fractions. These studies confirm lower contamination levels in fractions usually directly intended for human consumption (such as flour or semolina) and higher contamination levels in fractions intended for animal feed, such as bran and middlings. The concentrations in some fractions may be up to 800%, but more often range from 150% to 340%.[22]

Pascale *et al.* investigated the effects of cleaning in 10 samples of durum wheat contaminated with T-2 and HT-2 toxins by using a method based on liquid chromatography/mass spectrometry coupled with immunoaffinity column cleanup.[23] The method was validated in-house for the simultaneous analysis of both toxins, observing an overall reduction ranging from 11% to 89% and confirming the findings of a previous study conducted by Duarte *et al.*,[24] as well the reduction observed for the mycotoxin DON in semolina by Thammawong *et al.*,[25] even if the decrease of DON did typically range from 50% to 70%. The retention levels in semolina depend on the variety of the wheat, the penetration degree of *Fusarium* moulds and the transfer of the mycotoxin to the inner part of the kernel.[26] The concentration of DON in the bran fraction was reported by several authors starting from the 1980s.[4,8,9,27,28] Then, Hazel and Patel stated that typical results showed that the flour has about half the DON concentration of the cleaned wheat, while the bran can have DON concentrations two or more times greater than the wheat.[29] This trend suggested that the fungal infection was stronger near the surface of the kernels,[30,31] where mycotoxins were produced rather than transported from the kernel surface to the interior and that peripheral tissue is the first part colonised by fungi.[32,33] High levels of both type B tri-chothecenes (DON, NIV and 3-Ac-DON and 15-Ac-DON) and type A tri-chothecenes (HT-2 and T-2 toxin) in waste fractions (*e.g.*, screenings, outer layers of bran, *etc.*) were detected by Lancova *et al.*, and a substantial part of DON in cleaned grains was located in the bran, while approximately 40% of its original content was still left in the flour fractions.[30]

Moving to masked/modified forms, the simultaneous occurrence of DON and its conjugate, deoxynivalenol-3-glucoside (DON-3-Glc), has recently been documented in infected wheat. For example, Kostelanska *et al.* studied the fractionation of DON-3-Glc and DON in milling fractions, observing that their trend was similar, and white flours contained only approximately 60% of the content of DON-3-Glc and DON in unprocessed wheat grains.[34] Similar findings have been reported by Simsek *et al.* in analogous research in which the fate of DON and DON-3-Glc during milling was monitored by liquid chromatography–tandem mass spectrometry and gas chromatography an-alysis, respectively.[35] An approximate reduction of 61.8% in the detected DON level of the flour compared to the whole wheat was reported, whereas a 23.7% decrease in DON-3-Glc between the whole wheat and the flour was observed but was not found to be significant.

Regarding FBs, the milling process may decontaminate the corn myco-toxin content[36] after the germ separation because of the higher FB con-tamination than the whole caryopsis.[37] The high contamination content in bran and animal meal was demonstrated by Broggi *et al.*, who found a three-times higher FB contamination level in germ and bran than in whole corn.[38] Vanara *et al.* reported different results, finding the germ to have the same FB content as the whole kernel.[39] At the same time, the animal meal was the most contaminated fraction, confirming the findings of previous stud-ies.[37,38] Similar results were reported by Generotti *et al.*, conducting a study on the fate of FBs during the industrial production of cornmeal, processing from uncleaned corn to pre-cooked cornmeal.[40] The results obtained for the by-products, especially middlings, showed a very high FB increase (up to 600%), pointing out the necessity for careful management of the use of these by-products as feed for livestock. The germ was not significantly different from the starting raw material. Katta *et al.* suggested that the higher FB concentrations in germ, bran and fines might be due to the location of the fungus in the tip cap and germ areas just beneath the pericarp.[41] Several authors demonstrated the decrease of FB content in maize flour and meal after dry milling compared to the FB content in the whole kernels; in particular, a range of reduction from 26% to 69% was reported.[15] Bennet *et al.* conducted a study on the fate of FBs during wet milling in contaminated corn, finding that mycotoxins were dissolved into the steep-water or distributed to the gluten, fibre and germ fractions, leaving no detectable amounts in the starch, with concentrations ranging from approximately 10% (germ) to 40% (gluten) of the concentration in the raw corn kernels.[42] A similar investigation was performed by Lauren and Ringrose following the fate of ZEN, DON and NIV through a commercial wet-milling plant.[43] The authors obtained analogous findings for the latter two mycotoxins: the highly soluble DON and NIV were found at high levels in the steep-water fractions, but at only low levels in the germ, fibre and gluten. The opposite situation was found for the insoluble ZEN as low concentrations occurred in the steep-water and much higher levels were found in the solid fractions. Table 5.2 summarises the effects of milling on (masked/modified) mycotoxins.

5.3 Extrusion

Extrusion cooking is one of the fastest-growing food processes in recent years because of its many advantages over other traditional treatments. In extrusion cooking, raw materials are passed through continuous processing equipment where they are compressed and sheared at elevated temperatures and pressures to produce several types of by-products, including breakfast cereals, snacks or animal feeds. Although the temperatures employed and pressures are quite elevated, residence time is not sufficiently long to completely destroy mycotoxins. The results reported for the effects of extrusion are in fact rather variable, depending on extrusion time, temperature and other factors. Decreases of 100%, 95% and 83% for FBs, AFs and ZEN, respectively, were reported during corn extrusion, while lower decreases were monitored for DON, ochratoxin (OTA) and moniliformin (MON), where maximum reductions were 55%, 40% and 30%, respectively.[36]

The behaviour of DON and ZEN during the extrusion of naturally contaminated maize flour and maize grits using pilot-scale equipment was re-examined by Scudamore *et al.*[44] Their findings showed the stability of DON and ZEN during extrusion cooking, differing from the results obtained in a previous study conducted by Ryu *et al.* on the reduction in ZEN during

Table 5.2 Effects of milling on (masked) mycotoxins.

Mycotoxin(s); matrix/commodity	Ref.	Food processing; behaviour/effects
All	1	Review
All	21	Mycotoxins are not completely destroyed
DON	2	Review
All	22	Concentration in some fractions from 150 to 340%
T-2/HT-2	23	Concentration in screening and bran, reduction in cleaned wheat and semolina
T-2/HT-2	24	Reduction
DON	25	Reduction in semolina
All	26	Influence of different factors on mycotoxin distribution
DON	27	Reduction in flour; concentration in bran
DON	28	Reduction in flour; concentration in bran
DON	8	Concentration in bran
DON	9	Reduction in flour
DON	4	Reduction in flour
DON	29	Concentration in bran
DON/NIV/T-2/HT-2/3/ 15-Ac-DONs/FUSX	30	Concentrations in bran; reduction in flour
DON	31	Concentration in bran
AFs/ZEN	32	Peripheral tissue is more contaminated
All	33	Peripheral tissue is more contaminated
DON/DON-3-Glc	34	40% reduction in flour
DON/DON-3-Glc	35	Reduction in flour
FBs	36	Reduction
FBs	37, 38	Concentration in germ
FB$_1$ (dry milling)	39	Concentration in animal meal and reduction in flour and meal
FBs	40	Concentration in middlings
ALL	41	Fungal localisation near pericarp
FBs (cleaning/dry milling)	15	Reduction 26–69%
FBs (wet milling)	42	Concentration in steep-water or gluten, fibre and germ
ZEN/DON/NIV (wet milling)	43	DON/NIV: concentration in steep-water; ZEN: concentration in solid fractions

extrusion to make corn grits.[45] In the same study of DON and ZEN, Scudamore *et al.* also monitored the behaviour of FBs during extrusion cooking, and the degree of reduction of FBs was observed to depend on different factors, such as the presence of additives, reducing sugars and sodium chloride.[44] A few years before, another study provided by De Girolamo *et al.* investigated the stability of FB$_1$ and FB$_2$ during the production of cornflakes from raw cornflour by extrusion and roasting, finding an approximately 60–70% reduction in FB content during the entire process.[46] In this case, only 30% of those losses were attributed to the extrusion step, where the material was subjected to 70–170 °C for 2–5 minutes.

In general, the studies available in the scientific literature indicate that the greatest reduction of FBs occurs at temperatures of 160 °C or more and in the presence of glucose.[47] Extrusion of corn grits with 10% added glucose resulted in a 75–85% decrease in FB_1 levels. Some degradation products may be formed during extrusion, including a small amount of hydrolysed FB_1 and somewhat higher amounts of *N*-deoxy-fructosyl fumonision B1 (NDF-FB1) in extruded grits. In these cases, the degree of reduction seemed to depend on the type and amount of sugar added, as found by Seefelder *et al.*[48] and Castelo *et al.*,[49] who also studied the effects of added sugars on FB levels in extruded corn grits.

In contrast, the extrusion of the grits with and without sugars resulted in the same reduction level (64–67%) for AFs, as confirmed in a study conducted by Lu *et al.* in which the reduction of AFs after extrusion and toasting, with and without sugars, in cornflakes ranged from 78% to 85%.[50]

Looking now more on the side of mechanical effects, Castelo *et al.* showed the different impacts of the type of extruder on the FB_1 reduction level.[51] The authors observed that the loss of FB_1 from corn grits at extruding temperatures between 140 and 160 °C was significantly higher when using an extruder equipped with a mixing screw (29–69% loss) than an extruder fitted with a non-mixing screw (13–54% loss). Pineiro *et al.* found a 70–90% loss of FB_1 and FB_2 when cornflour was extruded using a single screw extruder at temperatures between 150 and 180 °C.[52]

Finally, regarding DON, Wu *et al.* noted that the moisture content and the compression were factors in its reduction, and a correlation between residence time of DON in the extruder and its degradation was observed when screws without a compression factor were used.[53] Extrusion cooking seemed to have little effect on the DON in wheat,[44] although when contaminated wheat was soaked in 5% aqueous sodium bisulphite beforehand, a relevant reduction was observed.[54]

5.4 Frying

Jackson *et al.* examined the effects of frying on FBs in corn-based foods and observed that degradation started to occur above 180 °C, confirming the heat stability of FBs in many commercial food processes.[55] However, a loss of 67% of the FBs was caused by frying corn chips for 15 minutes at 190 °C. Frying of corn at high temperatures (>190 °C) decreases FB concentrations in foods, with the degree of reduction achieved dependent on the cooking time and temperature, recipe and other factors.[47]

A reduction in DON of approximately 20% with respect to the initial natural contamination was observed by Samar *et al.* during the frying process, depending on the frying temperature: the best reduction was obtained when the fried covers reached the home-made colour at the lowest of the assayed temperatures.[56] Table 5.3 summarises the effects of extrusion or frying on (masked) mycotoxins.

Table 5.3 Effects of extrusion or frying on (masked) mycotoxins.

Mycotoxin(s); matrix/commodity	Ref.	Food processing; behaviour/effects
All	36	Reduction
FBs/DON/ZEN	44	DON/ZEN: no modifications; FBs: reduction dependent on sugars or sodium chloride occurrence
ZEN	45	Reduction 65–83%
FBs	46	Reduction 60–70%
FBs	47	Reduction at 190 °C
FBs	48	Reduction dependent on sugar added
FBs	49	Reduction dependent on sugar added
AFs	50	Reduction 78–85%
FBs	51	Reduction dependent on kind of extruder (mixing or no mixing screw)
FBs	52	70%/90% of reduction when cornflour was extruded using a single screw extruder at 150/180 °C
DON	53	Correlation between residence time in the extruder and DON reduction
DON	54	Relevant reduction when wheat was soaked in 5% aqueous sodium bisulphite
FBs (frying)	55	Degradation starts above 180 °C
DON (frying)	56	No reduction

5.5 Baking

Some studies have focused on the combined effect of the baking process on mycotoxin content in bakery products (bread, small size loaves, cakes, biscuits, *etc.*), and some of them reported significant reductions. In particular, several authors investigated the fate of mycotoxins during the bread-making process because bread is a very important commodity as a source of carbohydrates, proteins and vitamins B and E. Vidal *et al.* studied the fate of DON, DON-3-Glc and OTA during bread making.[57] DON increased from the unkneaded mix to fermented dough and decreased during baking, depending on the initial concentration in the flour. These observations were also in agreement with other previous studies, such as that performed by Wolff *et al.*, in which DON decreased up to 25% in baked bread.[58] Baking time seemed to have a much more important effect than baking temperatures on DON stability because the maximum temperature reached in the centre of the crumb was independent of the oven temperature. DON-3-Glc content increased both during kneading and fermentation as suggested by Zachariasova *et al.*, and also during baking, most likely due to the glycosylation of DON in the initial stages of baking before enzyme inactivation.[59] These results are in contrast with previous investigations. Kostelanska *et al.* noted that when so-called bread improver enzyme mixtures were employed as a dough ingredient, a distinct increase (up to 145%) of conjugated DON-3-Glc occurred in fermented dough.[34] On the other hand, some decrease in both DON-3-Glc and DON (10% and 13%, respectively, compared to fermented dough, and mainly in the crust) took place during baking. A similar trend was observed in the past[4] or recently in the study of Simsek *et al.*[35] DON levels detected during the fermentation stage were significantly higher in the mixed dough, and the baked bread had less DON-3-Glc detected than the dough. DON detection levels were higher after treatment with protease (16%) and xylanase (39%) compared to the wheat composite: the significant increases in apparent xylanase activity could cause the hydrolysis of cell wall material in the dough, resulting in a release of the bound DON and an increase in the DON concentration in the baking process samples.

The same trend observed by Vidal *et al.* on DON during bread making[57] was found by Bergamini *et al.*, studying how DON concentration may be influenced by modifying bread-making parameters.[60] Special emphasis was laid on the fermentation and baking stages, exploiting the power of a design of experiments approach to consider only those modifications that can really be applied on an industrial scale to obtain a final product that could be considered acceptable by consumers. The authors supposed that the DON increase during fermentation could be due to the enzymatic release of native DON from some bound forms occurring in the raw material. Studies at fermentation temperatures higher than 30 °C reported a reduction in DON concentrations from kneaded dough to fermented dough.[61]

A similar approach was employed by Suman *et al.* at the pilot plant/industrial scale to assess the effects of the changes in five technological parameters (fermentation time and temperature, baking time and temperature and presence of sodium bicarbonate) during fermentation and baking on the concentration of DON in crackers.[62] The results showed that the evolution of the toxin was significantly influenced by the baking temperature and time, while the other parameters seemed to have a smaller influence, most likely because of possible thermal degradation or rebinding with matrix constituents. The presence and the content of sodium bicarbonate suggested an opposite impact on DON concentration, due to a probable effect related to the pH conditions in the dough, which could modify the reactivity of the bound forms of the toxin.

Regarding baking, a wide range of results have been observed in different studies focusing on the changes in DON during the process. Some authors documented the high DON heat stability during baking at the temperatures of 170–350 °C, with no reduction of DON concentration after 30 minutes at 170 °C.[8,63,64] Abbas *et al.* reported a minimum decrease in DON concentration of 16.8%,[27] and Pacin *et al.* showed a mean of 42.3% less DON contamination between flour and bread.[65] Similar results were obtained by Neira *et al.*, reporting a maximum decrease of 96.6% lower DON in final products when compared to the dough, but in this case, no positive correlation between starting contamination level and reduction degree was noted.[66] More recently, another investigation on bread executed by Voss and Snook reported DON concentration reductions of up to more than 50% as a consequence of a combination of loss/degradation and dilution by recipe ingredients in bread.[67] At the same time, other studies highlighted practically no changes in DON concentrations, as in the investigation conducted by Lancova *et al.*, in which a DON increase due to kneading was reported, and its concentrations in bread baked at 210 °C for 14 minutes ranged from 94% to 100% compared with flour.[30] In a following study by Numanoglu *et al.*, after bread processing, again no significant reduction of DON was measured in the bread crust and crumb.[68] Studying the formation of degradation products of DON during food baking, different compounds were isolated and structural elucidation was achieved by nuclear magnetic resonance and mass spectrometry experiments. Kostelanska *et al.*, based on studies published earlier,[69,70] presumed that *nor*DON A, B, C, D, E, F and DON lactones might be the main degradation products of DON.[34] To confirm this assumption, a range of degradation experiments were conducted. Only some of these metabolites, presumably *nor*DON A, B and C, were found in thermally treated contaminated bread and only in crust, as expected, being exposed to higher temperatures. In the same experiment, five compounds named *nor*DON-3-Glc A, B, C, D and DON-3-Glc lactone were also identified and characterised for the very first time from DON-3-Glc thermal breakdown. In addition to the glucose-containing degradation products, *nor*DON A, B, C and D (typical for DON thermodegradation) were also detected in the model solution of heat-treated DON-3-Glc, indirectly indicating that the bond between DON and glucose could be cleaved during heat treatment. Preliminary indications are that thermal degradation of DON can be beneficial because *nor*DON A, B and C displayed no toxicity to preserved human kidney epithelial cells at a concentration of 100 µmol L^{-1}, whereas the EC$_{50}$ for DON toxicity in the same *in vitro* bioassay was 1.1 µmol L^{-1}. These findings are in agreement with the epoxy group playing an important role in the toxicity of trichothecenes, and the findings indicate that the degradation of DON under thermal treatment might therefore reduce the toxicity of DON-contaminated food.[69]

The recipe and the size of the loaf may be determinants for the calculation of the extent of mycotoxin reduction during baking. Boyaciouglu *et al.* studied the effect of additives, such as potassium bromate, L-ascorbic acid, sodium bisulphite, L-cysteine and ammonium phosphate on DON at different temperatures during baking.[71] They observed that while potassium bromate and L-ascorbic acid had no effect, L-cysteine and ammonium phosphate resulted in approximately 40% reductions, against the reduction of approximately 7% without additives.

Scudamore *et al.* manufactured bread, cakes and biscuits from flour contaminated with DON, ZEN and NIV.[72] These mycotoxins remained mostly unaffected during processing, although the reduction of DON during cake manufacture was greater than for bread because flours made up only approximately 25% of the starting ingredients. The stability of ZEN was observed in different studies conducted by Cano-Sancho *et al.* on its fate during the bread-making process.[73] The mycotoxin was stable after fermentation and baking under the experimental conditions that were used. ZEN is a very heat-stable compound.[74] Approximately 60% of the ZEN is estimated to survive after bread making and 80% after biscuit production.[75] Baking can substantially decrease ZEN levels, although given the thermal stability of the molecule, ZEN often survives the treatment as well.[76] Numanoglu *et al.* measured a maximum reduction of 25% after 15 minutes at 250 °C; in effect, the extent of ZEN reduction during thermal processes seems to be quite variable and dependent on the processing conditions applied.[77] Heating of NIV seemed to yield a mixture of four compounds (*nor*NIV A, *nor*NIV B, *nor*NIV C and NIV lactone), although only *nor*NIV B was detectable in a screening of several commercially available samples, most likely due to the very low contamination with free NIV itself. Cell culture experiments showed that the four compounds are less cytotoxic compared to NIV: whereas NIV showed an EC_{50} at 0.9 µmol, all of the other compounds failed to show any significant effect up to 100 µmol. Degradation of NIV was observed under all conditions, generally accelerating with increasing temperatures: after a heating time of 60 minutes, approximately 95% of the NIV was degraded, and under these conditions, *nor*NIV B was formed in the highest concentration compared to all other compounds.[78]

Valle-Algarra *et al.* studied the variation of the levels of OTA, NIV, 3-Ac-DON and DON during the bread-making process.[79] During the baking period, the average reductions, considering all the temperature–time conditions tested, were 32.9%, 76.9%, 65.6% and 47.9%, respectively, but they worked with OTA-spiked flour, which may be affected in a different way. Regarding T-2 and HT-2 toxins, Monaci *et al.* analysed the trends of these toxins in a bread model.[80] Most T-2 toxin was hydrolysed to HT-2 toxin during dough preparation, and approximately 20–30% of the HT-2 toxin was degraded during baking, disagreeing with a previous study conducted by Schwake-Anduschus *et al.*[13] where T-2 and HT-2 toxins were relatively stable during the baking process. The rate of the enzymatic conversion has recently been investigated in several cereal extracts, including maize, wheat, oats and barley, demonstrating that the speed of this conversion depends strongly on the type of cereal under investigation.[81] Although T-2 toxin was degraded under all conditions, generally accelerating with rising temperatures, only heating of T-2 toxin with α-D-glucose produced a mixture of three degradation products: HT-2, T-2 triol and T-2 tetraol. T-2 triol plays only a minor role in food samples, but the concentration levels of T-2 tetraol are often equal to or even higher than the concentration of T-2 toxin.[82] These T-2 toxin degradation products were found to be less cytotoxic compared to T-2 toxin, and their concentrations were estimated to be approximately 10–20% of the initial concentration of T-2 toxin. No significant difference between heating with or without water was detected.[83]

The baking process reduces the FB concentrations at temperatures above 150–200 °C in the final products, but the decrease most likely reflects chemical conversion of the FBs into other compounds and binding to sugars, proteins or other compounds occurring in the raw material or other recipe ingredients. For this reason, loss of FBs during baking may indicate that the FBs are extracted or removed by the products, destroyed, changed to other forms, bound to other components or rendered less extractable, so reduced levels are not directly indicative of reduced toxicity unless the removal from food has been unequivocally demonstrated.[47] The degree of reduction during baking depends on different factors, such as baking temperature and time, pH, moisture content and the recipe, particularly the type and amount of sugar. Twenty years ago, Scott *et al.* observed that by heating corn meal at 190 °C for 60 minutes, 60–80% of FB_1 and FB_2 were lost, while after baking at 220 °C for 25 minutes, the loss was almost total.[84] Another study conducted by Jackson *et al.* showed that decomposition of FB_1 and FB_2 began at 150 °C, and at temperatures above 175 °C, over 90% of the FBs were lost.[55] In contrast, Numanoglu *et al.* measured no significant reduction of FBs in the bread crust and crumb after bread processing.[68] Considering the mass balance of mycotoxins measured in flour and bread, only 3.1% of the total initial amounts of FBs were lost according to Castelo *et al.*, and reduction of FB_1 may be observed again only when high temperatures are used.[85] Table 5.4 summarises the effects on the levels of (masked/modified) mycotoxins during bread-making and baking processes.

Table 5.4 Effects on the levels of (masked) mycotoxins during bread-making and baking processes.

Mycotoxin(s); matrix/commodity	Ref.	Food processing; behaviour/effects
DON/DON-3-Glc/OTA (bread)	57	OTA: stable; DON: increase during fermentation; decrease during baking; DON-3-Glc: increase during fermentation and baking
DON (bread)	58	Reduction 25%
DON/DON-3-Glc (bread)	59	DON-3-Glc release
DON/DON-3-Glc (bread)	34	Stable during fermentation (DON-3-Glc increase up to 145% with improvers); reduction during baking
DON/DON-3-Glc (bread)	35	DON: increase during fermentation; DON-3-Glc: decrease during baking depending on enzyme added
DON (cookies/bread)	4	Reduction 35% (cookies); increase during fermentation at 30 °C
DON (bread)	60	Increase during fermentation and decrease during baking
DON	56	Reduction
DON (crackers)	62	Reduction during baking
DON (bread)	63	No reduction
DON (bread)	64	No reduction
DON (bread)	8	No reduction
DON	27	Reduction 16.8%
DON	65	Reduction 40%
DON (bread)	66	96.6% of reduction; no positive correlation between starting contamination level and reduction

Table 5.4 (*Continued*)

Mycotoxin(s); matrix/commodity	Ref.	Food processing; behaviour/effects
DON	67	Reduction from 61% (cookies) to 111% (pretzel); reduction 50% in donuts and bread.
DON/NIV/T-2/HT-2/3/ 15-Ac-DON/FUSX	30	No DON reduction at 210 °C for 14 minutes
DON, ZEN, FBs	68	No reduction
*nor*DONs	69	Identification of *nor*DON A–F and DON-lactones
*nor*DONs	70	Identification of *nor*DON A–F and DON-lactones
DON (bread)	71	Reduction 38%/46% with L-cysteine
DON/ZEN/NIV	72	No relevant reduction
DON/ZEN	73	No reduction
ZEN	74	No reduction
ZEN (bread/biscuits)	75	60% reduction in bread; 80% reduction in biscuits
ZEN	76	No reduction
ZEN (bread)	77	Maximum reduction (28%) at 250 °C for 15 minutes
NIV (degradation products)	78	*nor*NIV A–C + NIV-lactone: after 60 minutes, 95% of NIV transformed into *nor*NIV B
DON/3-Ac-DON/NIV/ OTA (bread)	79	DON: stable during fermentation and reduction during baking (47.9%); OTA: 30% reduction; Others: stable during fermentation and reduction during baking (no correlation time/ temperature)
DON/T-2/HT-2	80	T-2 hydrolysed into HT-2 during making; DON and HT-2 reduction (20%/30%) during baking.
T-2/HT-2	13	No reduction
T-2/HT-2	81	T-2/HT-2 conversion speed depending on kind of cereal
T-2 (degradation products)	82	Levels of T-2 tetraol are often equal or higher than T-2
T-2 (degradation products)	83	Occurrence at 200 °C for 1 hour (with or without water)
FBs	47	Reduction during baking at 190 °C
FBs	3	Reduction of 60%/80% at 190 °C for 60 minutes; complete reduction at 220 °C for 25 minutes
FBs	55	Degradation starts at 150 °C; 90% of reduction at 175 °C for 60 minutes (regardless of pH)
FB_1	85	Reduction at high temperatures

5.6 Pasta Manufacturing

Due to the high pasta consumption rates that characterise dietary habits, different studies on DON distribution during pasta-making and cooking processes were carried out, and the majority of them have reported reductions in DON levels in pasta ranging between 40% and 70%.[86,87] Brera *et al.* observed DON reductions with respect to semolina in dry and cooked pasta of 8% and 41%, respectively.[88] The stronger reduction observed for

Table 5.5 Effects on the levels of (masked) mycotoxins during the production of pasta/noodles.

Mycotoxin(s); matrix/commodity	Ref.	Food processing; behaviour/effects
DON	86	Reduction from 40 to 70%
DON	87	Reduction from 40 to 70%
DON	88	Reduction in cleaned wheat (39%), pealed wheat (66%) and semolina (63%); concentration in by-products
DON	89	Reduction in pasta and concentration in cooking water
DON	73	Reduction during boiling (75%); concentration in boiling water

cooked pasta was attributed to the high solubility of DON in water.[89] In this last work, the authors reported a consistent reduction of DON levels during each of the processing steps from uncleaned durum wheat to cooked pasta of approximately 25%, with a partition between cooked spaghetti and cooking water generally in favour of the cooking water. This trend was also confirmed by Cano-Sancho *et al.*[73] DON in pasta showed a clear decrease during boiling, reaching reduction values of up to 75% of the initial content after 10 minutes of boiling. In contrast, a direct increase in DON levels was reached in boiling water, of almost 75% of the initial concentration in pasta. Table 5.5 summarises the effects on the levels of mycotoxins during the production of pasta and noodles.

5.7 Tortilla Manufacturing

Another important process is alkali cooking or nixtamalisation. Nixtamalisation is used to produce snacks and tortillas and consists of first cooking the corn in alkaline water for a short period of time and then steeping the corn overnight.

Under alkaline conditions, FBs in contaminated corn are converted into the so-called hydrolysed FBs (HFBs), an aminopentol moiety formed by hydrolytically removing the two tricarballylic acid side chains from the 20-carbon FB backbone. FBs are water soluble and nixtamalisation lowers the FB content of food products if the cooking liquid is discarded.[47] Dombrink-Kurtzman *et al.* showed that nixtamalisation reduced the FB_1 concentration in tortillas by 81.5% and that FB_1 and HFB_1 were found mainly in the steeping and washing water.[90] Using the traditional nixtamalisation method of Mayan communities, the total FBs ($HFB_1 + FB_1$) in tortillas were reduced by 50%.

The residual lime and washing water also contained 50% of the total FB_1 in the starting material.[91] Differences exist between nixtamalisation as practiced in the home or other small-scale situations and as done in a large-scale industrial setting. Voss *et al.* investigated the production of fried

Table 5.6 Effects on the levels of (masked) mycotoxins during the production of tortillas.

Mycotoxin(s); matrix/commodity	Ref.	Food processing; behaviour/effects
FBs	47	Reduction during nixtamalisation
FB_1	90	Reduction of 81.5% during nixtamalisation (FB_1 and HFB_1 in steeping and washing water)
FBs	91	50% of the total FB_1 from the starting material in residual lime and washing water
FBs	92	FB_1 reduction up to 80%

tortilla chips in a pilot production line, and an FB_1 reduction of up to 80% compared to the raw corn was observed.[92] Chips contained HFB_1 at low concentrations, and almost no *N*-(carboxymethyl) fumonisin B1 (NCM-FB1) and NDF-FB_1 were detected. Cooking and steeping the corn in water was the critical step for reducing FBs in the masa and chips. Table 5.6 summarises the effects on the levels of (masked) mycotoxins during the production of tortillas.

5.8 Beer Production

The two main processes within beer making are malting and brewing. During malting, barley is steeped to obtain the correct moisture content and allowed to germinate to produce malt. The malt is then soaked in brewing liquor (mashing), and the temperature is raised to produce the wort that is then separated from the spent grains.

Scott *et al.* showed that DON was stable when wort was fermented with three different strains of *Saccharomyces cerevisiae* to produce beer, and whole ZEN was metabolised to form 69% β-zearalenol and 8% α-zearalenol.[93]

Great attention has been paid to a possible transfer of trichothecenes from the starting commodities barley and malt into the final beer products, and this concern was escalated by the finding of a substantial increase in masked DON-3-Glc across the beer-making chain.[30,34] Kostelanska *et al.* studied the fate of DON and its masked form DON-3-Glc during brewing processes, and their results documented the key role of the nature of the starting malt contamination.[94] The highest increases in DON and DON-3-Glc levels were even up to 536% and 210%, respectively, during the first monitoring period. When comparing input raw material and final product, fairly different trends were observed when different malts were used for identical techno-logical processing (in some cases, a decrease of DON and only a small in-crease of DON-3-Glc occurred). Mashing is the key phase influencing the mycotoxin concentration dynamics during the brewing processes. Under certain conditions, technological parameters such as the water-to-raw ma-terial ratio, temperature and enzymatic activity were shown to increase the DON concentrations dramatically (levels as high as 600% of the DON con-tent in the malt were reported).[95] A release of bound toxins can be assumed to take place under the conditions of mashing (malt suspended in water held

Table 5.7 Effects on the levels of (masked) mycotoxins during the production of beer.

Mycotoxin(s); matrix/commodity	Ref.	Food processing; behaviour/effects
DON/ZEN	93	DON: no reduction; ZEN: metabolised to form 69% β-zearalenol and 8% α-zearalenol
DON/DON-3-Glc	34	DON-3-Glc: increase; DON: depending on malt; no correlation between starting contamination level and final concentration; no correlation between alcohol and final concentration
DON/DON-3-Glc	30	DON reduction during steeping step, whereas DON increases during germination up to 250%
DON/DON-3-Glc	94	Key role of the nature of the starting malt contamination
DON	95	DON increases depending on different technological parameters
DON/DON-3-Glc	96	All extractable mycotoxins in malt transferred into the wort

at temperatures of approximately 40 °C for several hours) during physico-chemical and enzymatic processes. In spent grains, no DON and DON-3-Glc were detected, assuming that all the extractable mycotoxins contained in malt were effectively transferred into the wort.[96]

The earlier studies of Kostelanska *et al.*[34] and Lancova *et al.*[30] found that the levels of DON tended to decrease significantly during the steeping step, whereas germination was reported to increase the DON content in green barley by up to 250% of the amount present in malted barley, most likely due to additional mould growth and toxin production. Table 5.7 summarises the effects on the levels of (masked) mycotoxins during the production of beer.

5.9   Conclusion

The occurrence of mycotoxins in cereal-based foods and feeds is a global issue of significant concern due to their potential health risks for humans and livestock. In recent years, it has been shown that in foods contaminated by mycotoxins, the parent compounds may co-exist with compounds that are structurally related to the parent compounds and are generated through the metabolism of the plant or by the technological processes of food production.

Mycotoxins have always been considered to be thermostable compounds. However, interest in understanding the effects of food processing or cooking on mycotoxin levels has increased during the last decade. The technological processes employed seem to play an important role in the phenomenon associated with 'masking'. The mechanical energy of heat during the transformation process can cause significant changes, which can prompt reactions with macromolecules, such as sugars, proteins or lipids, or the release of the parent compound of the toxin after decomposition of the conjugated compound. Many factors must be considered in this research because complex physicochemical modifications occur throughout the transformation of raw ingredients into the final product. Recent evidence suggests that some food production processes lead to the reduction of levels of parent mycotoxins in finished products compared to the corresponding raw materials and ingredients, especially when high temperature conditions are involved: DON, ZEN or T-2 toxin mitigation during the baking of cereals-based products are real cases reported in the scientific literature. Another concrete example is represented for FBs, where the greatest reduction occurs at temperatures of 160 °C or more, while also dependent on pH (such as in the case of 'nixtamalisation'/alkaline water cooking processes), pressure, the presence of specific additives (such as ascorbic acid, amino acids and raising agents) and moisture content.

In some other circumstances, processing may stimulate the release of parent compounds from the conjugated forms: for instance, DON seems to increase from the unkneaded mix to fermented dough, and a possible explanation is here related to the enzymatic release of the native toxin from some bound forms occurring in the raw material during the fermentation steps.

To conclude, it is reasonable to consider that a significant effect of food processing is to reduce the overall toxicity of the finished products, as there are scientific indications that attest to a lower toxicity of the compounds obtained within the processing of foods than their precursors in raw materials. Nevertheless, there are large areas to explore in the coming years: one direction relates to the actual influence of the many variants of technological treatments that the modern food industry is able to employ. An important research area that will be deeply investigated in the future is the possible exploitation as mitigation strategies of those mycotoxin modifications occurring upon processing. This of course implies the understanding of the mechanisms leading to a decrease toxicity and bioavailability in the gastrointestinal tract of humans and livestock.

References

1. F. Cheli, L. Pinotti, L. Rossi and V. Dell'Orto, Effect of milling procedures on mycotoxin distribution in wheat fractions: a review, *LWT – Food Sci. Technol.*, 2013, **54**, 307–314.
2. M. Kushiro, Effects of milling and cooking processes on the deoxynivalenol content in wheat, *Int. J. Mol. Sci.*, 2008, **9**, 2127–2145.
3. P. M. Scott, S. R. Kanhere, J. E. Dexter, P. W. Brennan and H. L. Trenholm, Distribution of the trichothecene mycotoxin

deoxynivalenol (vomitoxin) during the milling of naturally contaminated hard red spring wheat and its fate in baked products, *J. Agric. Food Chem.*, 1984, **1**, 313–323.

4. J. C. Young, R. G. Fulcher, J. H. Hayhoe, P. M. Scott and J. E. Dexter, Effect of milling and baking on deoxynivalenol (vomitoxin) content of eastern Canadian wheats, *J. Agric. Food Chem.*, 1984, **32**, 659–664.

5. S. G. Edwards, E. T. Dickin, S. MacDonald, D. Buttler, C. M. Hazel, S. Patel and K. A. Scudamore, Distribution of *Fusarium* mycotoxins in UK wheat mill fractions, *Food Addit. Contam.*, 2011, **28**, 1694–1704.

6. T. Neuhof, M. Koch, T. Rasenko and I. Nehls, Occurrence of zearalenone in wheat kernels infected with *Fusarium culmorum*, *World Mycotoxin J.*, 2008, **1**, 429–435.

7. J. E. Dexter, B. A. Marchylo, R. M. Clear and J. M. Clarke, Effect of *Fusarium* head blight on semolina milling and pasta-making quality of durum wheat, *Cereal Chem.*, 1997, **74**, 519–525.

8. P. M. Scott, S. R. Kanhere, P.-Y. Lau, J. E. Dexter and R. Greenhalgh, Effects of experimental flour milling and bread baking on retention of deoxynivalenol (vomitoxin) in in hard red spring wheat, *Cereal Chem.*, 1983, **60**, 421–424.

9. L. M. Seitz, W. T. Yamazaki, R. L. Clements, H. E. Mohr and L. Andrews, Distribution of deoxynivalenol in soft wheat mill streams, *Cereal Chem.*, 1985, **62**, 467–469.

10. K. Tanaka, N. Hara, T. Goto and M. Manabe, Reduction of mycotoxins contamination by processing grain, *Proc. Int. Symp. Mycotoxicol.*, 1999, 95–100.

11. R. Tkachuk, J. E. Dexter, K. H. Tipples and T. W. Nowicki, Removal by specific gravity table of tombstone kernels and associated trichothecene from wheat infected with Fusarium head blight, *Cereal Chem.*, 1991, **68**, 428–431.

12. D. M. Trigo-Stockli, C. W. Deyoe, R. F. Satumbaga and J. R. Pedersen, Distribution of deoxynivalenol and zearalenone in milled fractions of wheat, *Cereal Chem.*, 1996, **73**, 388–391.

13. C. Schwake-Anduschus, G. Langenkämper, G. Unbehend, R. Dietrich, E. Märtlbauer and K. Münzing, Occurrence of Fusarium T-2 and HT-2 toxins in oats from cultivar studies in Germany and degradation of the toxins during grain cleaning treatment and food processing, *Food Addit. Contam.*, 2010, **27**(9), 1253–1260.

14. D. Saunders, F. Meredith and K. Voss, Control of fumonisin: effect of processing, *Environ. Health Perspect.*, 2001, **109**, 333–336.

15. E. W. Sydenham, L. Van der Westhuizen, S. Stockenström, G. S. Shepard and P. G. Thiel, Fumonisin-contaminated maize: physical treatment for the partial decontamination of bulk shipment, *Food Addit. Contam.*, 1994, **11**, 25–32.

16. K. A. Scudamore and S. Patel, Survey for aflatoxins, ochratoxin A, zearalenone and fumonisins in maize imported into the United Kingdom, *Food Addit. Contam.*, 2000, **17**, 407–416.

17. G. Aureli and M. G. D'Egidio, Efficacy of debranning on lowering of deoxynivalenol (DON) in manufacturing processes of durum wheat, *Tec. Molitoria*, 2007, **58**, 729–733.

18. F. Cheli, A. Campagnoli, V. Ventura, C. Brera, C. Berdini, E. Palmaccio and V. Dell'Orto, Effects of industrial processing on the distributions of deoxynivalenol, cadmium and lead in durum wheat milling fractions, *LWT – Food Sci. Technol.*, 2010, **43**, 1050–1057.

19. V. Sovrani, M. Blandino, V. Scarpino, A. Reyneri, J. D. Coïsson, F. Travaglia, M. Locatelli, M. Bordiga, R. Montella and M. Arlorio, Bioactive compound content, antioxidant activity, deoxynivalenol and heavy metal contamination of pearled wheat fractions, *Food Chem.*, 2012, **135**, 39–46.

20. G. Rios, L. Pinson-Gadais, J. Abecassis, N. Zakhia-Rozis and V. Lullien-Pellerin, Assessment of dehulling efficiency to reduce deoxynivalenol and *Fusarium* level in durum wheat grains, *J. Cereal Sci.*, 2009, **49**, 387–392.

21. T. Kuiper-Goodman, in Risk assessment and risk management of mycotoxins in food, *Mycotoxins in Food*, ed. N. Megan and M. Olsen, Cambridge, England, Woodhead, 2004, p. 3.

22. M. Barajas-Aceves, M. Hassan, R. Tinoco and R. Vazquez-Duhalt, Effect of pollutants on the ergosterol content as indicator of fungal biomass, *J. Microbiol. Methods*, 2002, **50**, 227–236.

23. M. Pascale, M. Haidukowski, V. M. T. Lattanzio, M. Silvestri, R. Ranieri and A. Visconti, Distribution of T-2 and HT-2 toxins in millling fractions of durum wheat, *J. Food Prot.*, 2011, 74(10), 1700–1707.

24. S. C. Duarte, A. Pena and C. M. Lino, A review on ochratoxin A occurrence and effects of processing of cereal and cereal derived food products, *Food Microbiol.*, 2010, **27**, 187–198.

25. N. Thammawong, H. Okadome, T. Shiina, H. Nakagawa, H. Nagashima, T. Nakajima and M. Kushiro, Distinct distribution of deoxynivalenol, nivalenol, and ergosterol in Fusarium-infected Japanese soft red winter wheat milling fractions, *Mycopathologia*, 2011, **172**, 323–330.

26. L. Pinson-Gadais, C. Barreau, M. Chaurand, S. Gregoire, M. Monmarson and F. Richard-Forget, Distribution of toxigenic *Fusarium* spp. and mycotoxin production in milling fractions of durum wheat, *Food Addit. Contam.*, 2007, **24**, 53–62.

27. H. K. Abbas, C. J. Mirocha, R. J. Pawlosky and D. J. Pusch, Effect of cleaning, milling, and baking on deoxynivalenol in wheat, *Appl. Environ. Microbiol.*, 1985, **50**, 482–486.

28. T. Tanaka, A. Hasegawa, S. Yamamoto, Y. Matsuki and Y. Ueno, Residues of Fusarium mycotoxins, nivalenol, deoxynivalenol and zearalenone, in wheat and processed food after milling and baking, *J. Food Hyg. Soc. Jpn.*, 1986, **27**, 653–655.

29. C. M. Hazel and S. Patel, Influence of processing on trichothecene levels, *Toxicol. Lett.*, 2004, **153**, 51–59.

30. K. Lancova, J. Hajslova, M. Kostelanska, J. Kohoutkova, J. Nedelnik, H. Moravcova and M. Vanova, Fate of trichothecene mycotoxins during

the processing: milling and baking, *Food Addit. Contam.*, 2008, **25**(5), 650–659.

31. Z. Nishio, K. Takata, M. Ito, M. Tanio, T. Tabiki, H. Yamauchi *et al.*, Deoxynivalenol distribution in flour and bran of spring wheat lines with different levels of *Fusarium* head blight resistance, *Plant Dis.*, 2010, **94**, 335–338.
32. C. Brera, C. Catano, B. De Santis, F. Debegnach, M. De Giacomo, E. Pannunzi and M. Miraglia, Effects of industrial processing on the distribution of aflatoxins and zearalenone in corn-milling fractions, *J. Agric. Food Chem.*, 2006, **54**, 5014–5019.
33. Y. Hemery, X. Rouau, V. Lullien-Pellerin, C. Barron and J. Abecassis, Dry processes to develop wheat fractions and products with enhanced nutritional quality, *J. Cereal Sci.*, 2007, **46**, 327–347.
34. M. Kostelanska, Z. Dzuman, A. Malachova, I. Capouchova, E. Prokinova, A. Skerikova *et al.*, Effects of milling and baking technologies on levels of deoxynivalenol and its masked form deoxynivalenol-3-glucoside, *J. Agric. Food Chem.*, 2011a, **59**(17), 9303–9312.
35. S. Simsek, K. Burgess, K. L. Whitney, Y. Gu and S. Y. Qian, Analysis of deoxynivalenol and deoxynivalenol-3-glucoside in wheat, *Food Control*, 2012, **26**, 287–292.
36. M. Castells, S. Marin, V. Sanchis and J. Ramos, Fate of mycotoxins in cereals during extrusion cooking: a review, *Food Addit. Contam.*, 2005, **22**(2), 150–157.
37. C. Brera, F. Debegnach, S. Grossi and M. Miraglia, Effect of the industrial processing on the distribution of fumonisin B1 in dry milling corn fractions, *J. Food Prot.*, 2004, **67**(6), 1261–1266.
38. L. E. Broggi, S. L. Resnik, A. M. Pacin, H. H. L. Gonzalez, G. Cano and D. Taglieri, Distribution of fumonisins in dry-milled corn fractions in Argentina, *Food Addit. Contam.*, 2002, **19**(5), 465–469.
39. F. Vanara, A. Reyneri and M. Blandino, Fate of fumonisin B1 in the processing of whole maize kernels during dry-milling, *Food Control*, 2009, **20**, 235–238.
40. S. Generotti, M. Cirlini, C. Dall'Asta and M. Suman, Influence of the industrial process from caryopsis to cornmeal semolina on levels of fumonisins and their masked forms, *Food Control*, 2015, **48**, 170–174.
41. S. K. Katta, A. E. Caganpang, L. S. Jackson and L. B. Bullerman, Distribution of *Fusarium* molds and fumonisins in dry-milled corn fractions, *Cereal Chem.*, 1997, **74**(6), 858–863.
42. G. A. Bennet, J. L. Richard and S. R. Eckhoff, Distribution of fumonisins in food and feed products prepared from contaminated corn, *Adv. Exp. Med. Biol.*, 1996, **392**, 317–322.
43. D. R. Lauren and M. A. Ringrose, Determination of the fate of three *Fusarium* mycotoxins through wet-milling of maize using an improved HPLC analytical technique, *Food Addit. Contam.*, 1997, **14**, 435–443.

44. K. A. Scudamore, R. C. E. Guy, B. Kelleher and S. J. MacDonald, Fate of *Fusarium* mycotoxins in maize flour and grits during extrusion cooking, *Food Addit. Contam.*, 2008, **25**(11), 1374–1384.

45. D. Ryu, M. A. Hanna and L. B. Bullerman, Stability of zearalenone during extrusion of corn grits, *J. Food Prot.*, 1999, **62**, 1482–1484.

46. A. De Girolamo, M. Solfrizzo and A. Visconti, Effect of processing on fumonisin concentration in corn flakes, *J. Food Prot.*, 2001, **64**(5), 701–705.

47. H. U. Humpf and K. A. Voss, Effects of thermal food processing on the chemical structure and toxicity of fumonisin mycotoxins, *Mol. Nutr. Food Res.*, 2004, **48**, 255–269.

48. W. Seefelder, M. Hartl and H. U. Humpf, Determination of N-(carboxymethyl)fumonisin B1 in corn products by liquid chromatography/electrospray ionization mass spectrometry, *J. Agric. Food Chem.*, 2001, **49**, 2146–2151.

49. M. M. Castelo, L. S. Jackson, M. A. Hanna, B. H. Reynolds and L. B. Bullerman, Loss of fumonisin B1 in extruded and baked corn-based foods with sugars, *J. Food Sci.*, 2001, **66**, 416–421.

50. Z. Lu, W. R. Dantzer, E. C. Hopmans, V. Prisk, J. E. Cunnick, P. A. Murphy and S. Hendrich, Reaction with fructose detoxifies fumonisin B1 while stimulating liver associated natural killer cell activity in rats, *J. Agric. Food Chem.*, 1997, **45**, 803–809.

51. M. M. Castelo, S. K. Katta, S. S. Sumner, M. A. Hanna and L. B. Bullermann, Extrusion cooking reduces recoverability of fumonisin B1 from extruded corn grits, *J. Food Sci.*, 1998a, **63**, 696–698.

52. M. Pineiro, J. Miller, G. Silva and S. Musser, Effect of commercial processing on fumonisin concentrations of maize-based foods, *Mycotoxins Res.*, 1999, **15**, 2–12.

53. Q. Wu, L. Lohrey, B. Cramer, Z. Yuan and H. U. Humpf, Impact of physicochemical parameters on the decomposition of deoxynivalenol during extrusion cooking of wheat grits, *J. Agric. Food Chem.*, 2011, **29**, 12480–12485.

54. M. Accerbi, V. E. Rinadi and P. K. Ng, Utilization of highly deoxynivalenol-contaminated wheat via extrusion processing, *J. Food Protect.*, 1999, **62**, 1485–1487.

55. L. S. Jackson, J. J. Hlywka, K. R. Senthil and L. B. Bullerman, Effects of thermal processing on the stability of fumonisin B2 in an aqueous system, *J. Agric. Food Chem.*, 1996, **44**, 1984–1987.

56. M. M. Samar, S. L. Resnik, H. H. L. Gonzalez, A. M. Pacin and M. D. Castillo, Deoxynivalenol reduction during the frying process of turnover pie coveres, *Food Control*, 2007, **18**, 1295–1299.

57. A. Vidal, H. Morales, V. Sanchis, A. J. Ramos and S. Marìn, Stability of DON and OTA during the breadmaking process and determination of process and performance criteria, *Food Control*, 2014, **40**, 234–242.

58. J. Wolff, Untersuchungen der Gehaltsveränderungen der Fusariumtoxine Deoxynivalenol und ZENralenon durch Berund Verarbeitungsprozesse in

Getreide und Getreideprodukten, 26, Mykotoxin-Workshop, Herrsching, Germany, 2004.

59. M. Zachariasova, M. Vaclavikova, O. Lacina, L. Vaclavik and J. Hajslova, Deoxynivalenol oligoglycosides: new "masked" *Fusarium* toxins occurring in malt, beer, and breadstuff, *J. Agric. Food Chem.*, 2012, **60**(36), 9280–9291.

60. E. Bergamini, D. Catellani, C. Dall'Asta, G. Galaverna, A. Dossena and R. Marchelli, Fate of *Fusarium* mycotoxins in the cereal product supply chain: the deoxynivalenol (DON) case within industrial bread-making technology, *Food Addit. Contam.*, 2010, **27**, 677–687.

61. M. M. Samar, C. F. Fontàn, S. L. Resnik and A. Pacin, Effect of fermentation on naturally occurring deoxynivalenol (DON) in Argentina bread processing technology, *Food Addit. Contam.*, 2001, **18**, 1004–1010.

62. M. Suman, A. Manzitti and D. Catellani, A design of experiments approach to studying deoxynivalenol and deoxynivalenol-3-glucoside evolution throughout industrial production of wholegrain crackers exploiting LC-MS/MS technique, *World Mycotoxin J*, 2012, **5**(3), 241–249.

63. A. A. El-Banna, P.-Y. Lau and P. M. Scott, Fate of mycotoxins during processing of foodstuffs. II-Deoxynivalenol (vomitoxin) during making of Egyptian bread, *J. Food Prot.*, 1983, **46**, 484–486.

64. B. H. Gärtner, Munich M., Kleijer G., Mascher F. Characterisation of kernel resistance against Fusarium infection in spring wheat by baking quality and mycotoxin assessments, *Eur. J. Plant Pathol.*, 2008, **120**, 61–68.

65. A. Pacin, E. C. Bovier, G. Cano, D. Taglieri and C. H. Pezzani, Effect of the bread making process on wheat flour contaminated by deoxynivalenol and exposure estimate, *Food Control*, 2010, **21**, 492–495.

66. M. S. Neira, A. M. Pacin, E. J. Martìnez, G. Moltò and S. L. Resnik, The effects of bakery processing on naturally deoxynivalenol contamination, *Int. J. Food Microbiol.*, 1997, **37**(1), 21–25.

67. K. A. Voss and M. E. Snook, Stability of the mycotoxin deoxynivalenol (DON) during the production of flour-based foods and wheat flake cereal, *Food Addit. Contam.*, 2010, **27**(12), 1694–1700.

68. E. Numanoglu, U. Uygun, H. Koksel and M. Solfrizzo, Stability of *Fusarium* toxins during traditional Turkish maize bread production, *QAS*, 2010, **2**(2), 84–92.

69. M. Bretz, M. Beyer, B. Cramer, A. Knecht and H. U. Humpf, Thermal degradation of the *Fusarium* mycotoxin deoxynivalenol, *J. Agric. Food Chem.*, 2006, **54**, 6445–6451.

70. R. Greenhalgh, J. Gilbert, R. R. King, B. A. Blackwell, J. R. Startin and M. J. Shepherd, Synthesis, characterization, and occurrence in bread and cereal products of an isomer of 4-deoxynivalenol (vomitoxin), *J. Agric. Food Chem.*, 1984, **32**, 1416–1420.

71. D. Boyacioglu, N. S. Hettiarachchy and B. L. D'Appolonia, Additives affect deoxynivlenol (vomitoxin) flour during breadbaking, *J. Food Sci.*, 1993, **58**, 416–418.

72. K. A. Scudamore, C. M. Hazel, S. Patel and F. Scriven, Deoxynivalenol and other *Fusarium* mycotoxins in bread, cake and biscuits produced from UK-grown wheat under commercial and pilot scale conditions, *Food Addit. Contam.*, 2009, **26**(8), 1191–1198.

73. G. Cano-Sancho, V. Sanchis, A. J. Ramos and S. Marìn, Effect of food processing on exposure assessment studies with mycotoxins, *Food Addit. Contam.*, 2013, **30**(5), 867–875.

74. D. Ryu, M. A. Hanna, K. M. Eskridge and L. B. Bullerman, Heat stability of zearalenone in an aqueous buffered model system, *J. Agric. Food Chem.*, 2003, **51**, 1746–1748.

75. A. J. Alldrick, M. Hajšelová, *Mycotoxins in Food Detection and Control*, CRC Woodhead, Cambridge, ch. 15, ZENralenone, 2004, pp. 353–362.

76. C. M. Maragos, Zearalenone occurrence and human exposure, *World Mycotoxin J.*, 2010, **3**, 369–383.

77. E. Numanoglu, S. Yener, V. Gökmen, U. Uygun and H. Koksel, Modelling thermal degradation of zearalenone in maize bread during baking, *Food Addit. Contam.*, 2013, **30**(3), 528–533.

78. M. Bretz, A. Knecht, S. Göckler and H.-U. Humpf, Structural elucidation and analysis of thermal degradation products of the *Fusarium* mycotoxin nivalenol, *Mol. Nutr. Food Res.*, 2005, **49**, 309–316.

79. F. M. Valle-Algarra, E. M. Mateo, A. Medina, J. V. Gimeno-Adelantado and M. Jimenéz, Changes in ochratoxin A and type B trichothecenes contained in wheat flour during dough fermentation and bread baking processes, *Food Addit. Contam.*, 2009, **26**(6), 896–906.

80. L. Monaci, E. De Angelis and A. Visconti, Determination of deoxynivalenol, T-2 and HT-2 toxins in a bread model food by liquid chromatography-high resolution-Orbitrap-mass spectrometry equipped with a high-energy collision dissociation cell, *J. Chromatogr. A*, 2011, **1218**, 8646–8654.

81. V. M. T. Lattanzio, M. Solfrizzo and A. Visconti, Enzymatic hydrolysis of T-2 toxin for the quantitative determination of total T-2 and HT-2 toxins in cereals, *Anal. Bioanal. Chem.*, 2009, **395**, 1325–1334.

82. C. Gottschalk, J. Barthel, G. Engelhardt, J. Bauer and K. Meyer, Occurrence of type A trichothecenes in conventionally and organically produced oats and oat products, *Mol. Nutr. Food Res.*, 2007, **51**, 1547–1553.

83. M. Beyer, I. Ferse, D. Mulac, E.-U. Würthwein and H.-U. Humpf, Structural elucidation of T-2 toxin thermal degradation products and investigation toward their occurrence in retail food, *J. Agric. Food Chem.*, 2009, **57**, 1867–1875.

84. P. M. Scott and G. A. Lawrence, Stability and problems in recovery of fumonisins added to corn-based foods, *J. AOAC Int.*, 1994, 77, 541–545.

85. M. M. Castelo, S. S. Sumner and L. B. Bullerman, Stability of fumonisins in thermally processed corn products, *J. Food Prott.*, 1998b, **61**(8), 1030–1033.

86. T. W. Nowicki, D. G. Gaba, J. E. Dexter, R. R. Matsuo and R. M. Clear, Retention of the Fusarium mycotoxin deoxynivalenol in wheat during

processing and cooking of spaghetti and noodles, *J. Cereal Sci.*, 1988, **8**, 189–202.

87. Y. Sugita-Konishi, B. J. Park, K. Kobayashi-Hattori, T. Tanaka, T. Chonan, K. Yoshikawa and S. Kumagai, Effect of cooking process on the deoxynivalenol content and its subsequent cytotoxicity in wheat products, *Biosci., Biotechnol., Biochem.*, 2006, **70**, 1764–1768.

88. C. Brera, A. Peduto, F. Debegnach, E. Pannunzi, E. Prantera, E. Gregori, M. De Giacomo and B. De Santis, Study of the influence of the milling process on the distribution of deoxynivalenol content from the caryopsis to the cooked pasta, *Food Control*, 2013, **32**, 309–312.

89. A. Visconti, E. M. Haidukowski, M. Pascale and M. Silvestri, Reduction of deoxynivalenol during durum wheat processing and spaghetti cooking, *Toxicol. Lett.*, 2004, **153**, 181–189.

90. M. A. Dombrink-Kurtzman, T. J. Dvorak, M. E. Barron and L. W. Rooney, Effect of nixtamalization (alkaline cooking) on fumonisin-contaminated corn for production of masa and tortillas, *J. Agric. Food Chem.*, 2000, **48**, 5781–5786.

91. E. Palencia, O. Torres, W. Hogler and E. I. Meredith, Total fumonisins are reduced in tortillas using the traditional nixtamalization method of Mayan communities, *J. Nutr.*, 2003, **133**, 3200–3203.

92. K. A. Voss, S. M. Poling, E. I. Meredith, C. W. Bacon and D. S. Saunders, Fate of fumonisins during the production of fried tortilla chips, *J. Agric. Food Chem.*, 2001, **49**, 3120–3126.

93. P. M. Scott, S. R. Kanhere, E. F. Daley and J. M. Farber, Fermentation of wort containing deoxynivalenol and zearalenone, *Mycotoxin Res.*, 1992, **8**, 58–65.

94. M. Kostelanska, M. Zachariasova, O. Lacina, M. Fenclova, A. L. Kollos and J. Hajslova, The study of deoxynivalenol and its masked metabolites fate during the brewing process realised by UPLC-TOFMS method, *Food Chem.*, 2011b, **126**, 1870–1876.

95. L. Niessen, M. Bohm-Schrami, H. Vogel and S. Donhauser, Deoxynivalenol in commercial beer-Screening for the toxin with an indirect competitive ELISA, *Mycotoxin Res.*, 1993, **9**, 99–108.

96. M. Kostelanska, J. Hajšlová, M. Zachariasova, A. Malachova, K. Kalachova, J. Poustka, J. Fiala, P. M. Scott, F. Berthiller and R. Krska, Occurrence of deoxynivalenol and its major conjugate, deoxynivalenol-3-glucoside, in beer and some brewing intermediates, *J. Agric. Food Chem.*, 2009, **57**, 3187–3194.

In Vitro *Assays to Estimate the Toxicological Effects of Masked Mycotoxins*

ALEXIS V. NATHANAIL,*[a] MARIKA JESTOI,[b]
MARTINA JONSSON[a] AND KIMMO PELTONEN[c]

[a] Chemistry and Toxicology Unit, Research and Laboratory Department, Finnish Food Safety Authority (Evira), Mustialankatu 3, 00790 Helsinki, Finland; [b] Product Safety Unit, Control Department, Finnish Food Safety Authority (Evira), Mustialankatu 3, 00790 Helsinki, Finland; [c] Finnish Safety and Chemicals Agency (Tukes), Opastinsilta 12 B, 00521 Helsinki, Finland
*Email: alexis.nathanail@helsinki.fi

6.1 Introduction

Progress has been made to the present day in almost all research areas concerning masked mycotoxins, including their formation, analysis and occurrence, as well as the fates of these compounds during food manufacturing processes. However, toxicological information on masked mycotoxins is scarce, even though the literature is rich in reports concerning precursor mycotoxins' acute toxicity, cellular mechanisms, toxic manifestations and pathological implications in animal performance. Masked mycotoxins structurally differ from their parent compounds, a fact that may result in differences in their polarity, solubility and other chemical attributes, subsequently altering their toxicological properties (bioavailability, toxicokinetics, toxicodynamics and general toxicity).[1] Due to the natural

Issues in Toxicology No. 24
Masked Mycotoxins in Food: Formation, Occurrence and Toxicological Relevance
Edited by Chiara Dall'Asta and Franz Berthiller
© The Royal Society of Chemistry 2016
Published by the Royal Society of Chemistry, www.rsc.org

co-occurrence of different mycotoxins along with their derivatives, the toxicity of contaminated food or feed cannot be accurately estimated by determining the concentration of a single or a few toxins. Therefore, the natural co-occurrence of parent compounds with some other fungal metabolites and their masked forms must be taken into consideration during risk analysis. Another major concern, besides the possible inherent toxicity of some masked mycotoxins, is their potential to hydrolyse within the mammalian gastrointestinal tract (GIT), leading to an underestimation of the total exposure compared to that determined by conventional analytical approaches.

Considerable speculation surrounds the relevance of masked mycotoxins to human and/or animal health. Masked mycotoxins are not covered by existing legislation, and current regulatory limits are solely based on parent mycotoxins. Obtaining toxicological information on masked mycotoxins is a major obstacle in risk assessment, due to the limited availability of pure compounds that would allow comprehensive *in vivo* toxicity studies. The application of non-animal test methods, such as *in vitro* methodologies, provides toxicologists with powerful tools to enhance understanding of the hazardous effects of chemicals and predict the likelihood of health implications, with minimum substance requirements.[2] In this chapter, we aim to review all current toxicological knowledge that has been obtained using *in vitro* models, in the aspects of the toxicity and digestive fates of masked mycotoxins. In addition, the relevance of masked mycotoxins is discussed from a toxicological perspective and a selection of *in vitro* techniques is compiled that could generate important new information to better assess the risks associated with these compounds.

6.2   Modern Mycotoxicology: Masked Mycotoxins

Toxicology has been defined as the study of the adverse effects of xenobiotics on living organisms. Modern toxicology, goes beyond this definition and dives into the cellular, biochemical and molecular mechanisms associated with the action of exogenous agents, often by employing them as research tools.[3] Furthermore, modern toxicology investigates functional effects (*e.g.* neurobehavioural/immunological implications) and aims to assess the probability of their occurrence. Mycotoxins, as ubiquitous contaminants of the food chain, have been extensively studied over the past decades for their general and acute toxicity, but the current focus is turning more towards their mechanistic and (sub)chronic manifestations.[4] Nevertheless, several challenges exist in mycotoxicology (*i.e.* the study of fungal toxins and their adverse health effects), particularly with regard to food safety. Among these are the possible chemical interactions of mycotoxins with food components, the metabolic activation during digestion and the influence of toxicokinetic/toxicodynamic parameters on the observed toxicity. A cost–benefit analysis of food consumption and its nutritional value *versus* the simultaneous exposure to usually low concentrations of harmful substances has to be kept in mind as well.

6.2.1 Adverse Effects of Mycotoxins

Most of the known mycotoxins have a primary *in vivo* effect on a specific body system, and based on this, a crude approach has been implemented to classify them accordingly. In this context, mycotoxins may be considered as immunotoxic, haematotoxic, hepatotoxic, nephrotoxic, teratogenic, neurotoxic, mutagenic, carcinogenic or dermonecrotic, or they may induce toxicity to the reproductive systems.[5,6] This simplified classification does not necessarily cover all possible scenarios because, occasionally, several biological systems can be simultaneously affected by exposure to a single mycotoxin. Moreover, mycotoxin toxicity may either be inherent to the molecule or require metabolic activation within the exposed organism. This activation phenomenon is observed for instance in aflatoxin B_1 (AFB_1) and its ultimate mutagenic and carcinogenic metabolite, AFB_1-8,9-epoxide.[7] Additionally, the natural co-occurrence of fungal metabolites can result in synergistic, additive, potentiating or antagonistic interactions, which further complicates the toxicological categorisation of mycotoxins.

The critical factors underlying the biological activity of a compound include its physico-chemical properties, chemical structure, stereochemistry and the presence of active moieties in the molecule. In this regard, the toxicities of different mycotoxins vary greatly due to their chemical diversity.[8] As examples of the toxicological variation among mycotoxins, deoxynivalenol (DON), a type-B trichothecene, is known to inhibit protein synthesis by binding to ribosomes,[9] AFB_1 causes damage to DNA after metabolic activation,[7] whereas moniliformin, a low-molecular-weight *Fusarium* mycotoxin, leads to mitochondrial dysfunction.[10] On a biochemical level, the modes of action of mycotoxins can be divided into four categories: interactions with DNA, the inhibition of protein synthesis, effects on cell membranes and the disruption of energy metabolism (*e.g.* affecting the biosynthesis of ATP).

Another distinction between the types of adverse effects of xenobiotics, including mycotoxins, is based on the general site of action. Local effects are those that occur at the site of first contact between a toxicant and a biological system.[3] DON, for instance, is known to cause haemorrhaging of the upper GIT when in direct contact with intestinal epithelial cells.[11] Systemic effects, on the other hand, require the absorption and distribution of a toxicant from its entry point to a distant site at which toxic effects are produced. Most chemicals that exert systemic effects do not cause a similar degree of toxicity in all organs. Instead, they usually elicit their main toxicity in only one or two target organs. It should be noted that the target organ is seldom the site of the highest concentration of a toxicant.[3] The central nervous system, the circulatory system and the blood and haematopoietic system are most frequently involved in systemic toxicity, followed by visceral organs such as the liver, kidneys and lungs. DON, in addition to having local effects, can also enter the systemic circulation and cause deleterious effects on the immune, intestinal and neuroendocrine systems.[12]

Regarding the toxicological manifestations of mycotoxins, animal mycotoxicoses have been reported more often in the literature than acute mycotoxin-induced human diseases.[13,14] As attention is drawn more towards the role of mycotoxins in terms of (sub)chronic toxicology, both in human and in animal nutrition, new insights into their metabolic fate and their impact on health due to constant low exposure will be crucial in future risk assessments. These (sub)chronic effects include the involvement of mycotoxins as aetiological factors in different human diseases, as well as their ability to evoke feed refusal, reduced productivity, poorer reproductive capabilities or diminished resistance to infectious agents in animals.[6] The immunosuppressive actions of mycotoxins are of particular interest as they may be overshadowed by other, more easily recognisable symptoms.

6.2.2 'Masking' of Mycotoxins: Toxicological Repercussions

Organisms exploit an arsenal of defence mechanisms to counter the adverse effects of toxic substances. These mechanisms may instigate decreased exposure to a xenobiotic *via* toxicokinetically derived resistance (*e.g.* reduced uptake, biotransformation and increased elimination of toxicants) and/or decreased sensitivity to a xenobiotic *via* toxicodynamically derived resistance (*e.g.* toxin target receptor mutations and circumvention of inhibited pathways).[15] A very prominent strategy of toxicokinetically derived resistance is the biotransformation of toxicants (phase I and II metabolism), followed by either excretion in animals and humans or compartmentation (phase III) in plants.[16] During phase I metabolism, reactive or polar groups are introduced by enzymatic action onto usually lipophilic xenobiotics by oxidation, reduction and hydrolysis reactions. These reactions do not always lead to components with reduced toxicity; in some cases, the resulting metabolites can be even more toxic than the parent compounds. Phase II biotransformation reactions include glucuronidation, glucosylation (Glc), sulfation (S) and conjugation with glutathione (GSH) or amino acids. Conjugation reactions of toxicants with hydrophilic molecules give rise to end-products usually having an increased molecular weight and hydrophilicity, different chemical properties and occasionally unpredictable toxicological behaviour.[17] Modified mycotoxins can emerge through such metabolic biotransformations in living plants (masked mycotoxins), mammals and fungi, or can be formed during food manufacturing processes.[18] In plants, there is evidence that phase II glucosylated or glutathionylated biotransformation products are eliminated from the cytosol *via* ATP-binding cassette transporters into the vacuolar or apoplastic space.[19,20]

Several exposure routes exist for mycotoxins and their derivatives, the most important being the oral route, whereas the lungs or skin might only represent a possible route in certain occupational cases (*e.g.* farmers and bakers). As with most xenobiotics, absorption of masked mycotoxins may occur in different parts of the body, but mainly takes place in the upper and lower GIT. Exposure to masked mycotoxins may lead to reactivation, metabolism or potential adverse effects on a target organ due to inherent or parent toxicity. Parent toxicity is the result of de-conjugation of masked mycotoxins within the GIT, releasing the native toxins, which in turn can exert local or systemic effects depending on the toxicological profile of each toxicant. Figure 6.1 presents the theoretical fate of masked mycotoxins in monogastric species. Depending on the impact that 'masking' transformations may have on the bioavailability of a certain mycotoxin, its uptake and transport into the systemic circulation can be significantly affected. As soon as a masked mycotoxin is released from the food matrix, it may undergo biotransformation. Most enzymes and enzyme systems that catalyse these metabolic reactions are localised in the endoplasmic reticulum of liver cells.[21] These enzymes are also located at the main entry sites of xenobiotics into the body, such as the skin, lungs and GIT, as well as numerous other organs (*e.g.* kidneys and pancreas).

Although the toxicities of the parent compounds such as DON or zearalenone (ZEN) are well described in the literature,[11,12,22] the toxicological relevance of their masked forms is largely unexplored. In most masked mycotoxins that have been identified, the active sites of the molecules remain intact after conjugation reactions. Consequently, masked forms may at least in theory maintain their toxic potential. Nonetheless, the introduction of a conjugate residue to a xenobiotic produces a molecule with increased polarity that may hamper its ability to be absorbed through passive diffusion, the mechanism by which most mycotoxins are believed to pass across the intestinal cells.[23] Besides changes in toxicokinetic parameters of xenobiotics, conjugated moieties may affect the interaction of precursor molecules with target sites (toxicodynamics), leading to metabolites with decreased acute toxicity. Furthermore, the toxic effects that a released mycotoxin may exert on the gut microbiota are very important concerns that must be investigated. For all of these reasons, the classification of mycotoxin conjugates as masked or simply as detoxification products requires in-depth toxicological knowledge, as well as a thorough understanding of their fate during food processing and digestion.

6.3   Toxicity Testing *In Vitro*

Risk assessment of chemicals is currently shifting away from *in vivo* testing towards alternative (non-animal) approaches, not only due to the high cost of such experimental setups, but also on the grounds of public opinion and legislative changes. This includes the full marketing ban in the European Union of cosmetics containing new ingredients tested on animals,[24] as well as additional regulatory demands (*e.g.* Registration, Evaluation, Authorisation and Restriction of CHemicals [REACH]). To date, even though complex toxicological endpoints such as repeated dose toxicity, reproductive toxicity and toxicokinetics are still mainly determined under obligatory *in vivo* animal testing, there have been efforts in recent legislative

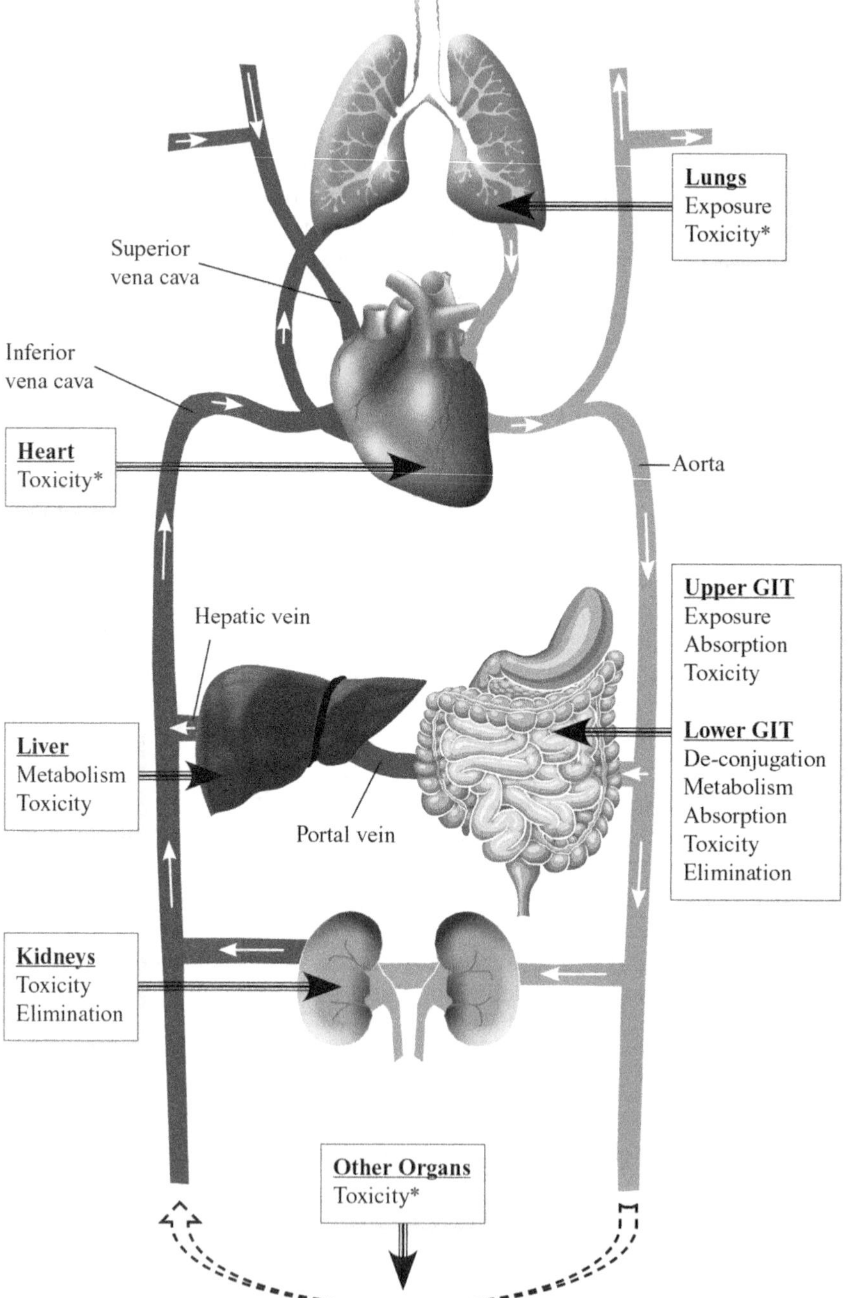

Figure 6.1 Theoretical fate of masked mycotoxins in monogastric species. *Unconfirmed.*

frameworks to replace them with alternative *in vitro* and/or computational *in silico* methods.[25] Alternative toxicological approaches are cost effective, less time consuming and require reduced amounts of test substances in comparison to animal experiments. Additionally, *in vitro* methods are usually easy to set up, simple to automate and offer satisfactory repeatability if properly tested and validated. Advances in science and technology have led to the development of novel tests based on human cells, or engineered tissues that can yield better results in terms of predicting potential toxic effects on humans without the need for laboratory animals.[26]

The term *in vitro* toxicology testing refers to the handling of cells and tissues outside of intact organ systems under conditions that support their growth, differentiation and stability. Since their discovery in the 1960s, these methods have proven essential for developing and performing various types of toxicological studies, earlier conducted exclusively *in vivo*. During the past decade, cell culture technologies have tremendously improved as a result of the scientific demand for rapid, simple and efficient methods for a broad array of applications.[27] *In vitro* methods are not only useful for assessment of the hazardous potential of chemicals in foodstuffs, but they can also be used to gain a mechanistic understanding of toxicologically important processes in experimental animals, as well as in humans.[27] *In vitro* test systems are especially well suited to investigating low-molecular-weight chemicals such as natural toxins, and also allow the critical assessment of complex mixtures to estimate the additive, synergistic or antagonistic effects of complex mixtures. These so-called cocktail effects are a high priority in risk assessment today.

Another advantage of *in vitro* toxicity tests is that they can be more easily validated, in contrast to most *in vivo* tests, which can be imprecise and outdated, even though internationally accepted standards have been established to improve their reliability. Complex animal experiments are prone to generating data that are difficult to interpret, mostly due to the lack of knowledge regarding the metabolic fate and internal distribution of test chemicals, problems of inter-species extrapolation and the use of high dosage levels. As a consequence, processes such as the replication of experiments, testing of toxicant mixtures and establishing of dose–response relationships are more difficult to handle under *in vivo* experimental conditions. Adequate consideration of food matrix components is also important and one of the main challenges in food toxicology is the investigation of food as a whole. *In vitro* assays are capable of providing extensive information on how food constituents interact with human cells and macromolecules. Consideration of the chemical structure might also suggest the requirement for modelling specific types of *in vivo* metabolism, such as those affected by intestinal enzymes and the intestinal microflora, or other potential extra-hepatic metabolisms, with ethanol to acetaldehyde metabolism in the gut being a good example. By focusing on those mechanisms that are relevant to humans, it should be possible to identify key toxicological responses *in vitro*.

In order to achieve accurate simulation of these parameters with an *in vitro* experimental design, realistic reproduction of all essential organs and processes involved is necessary. Furthermore, the effects of isolated or so-called 'pure' toxins, applied in the laboratory, are not easily generalised to real-life conditions, as several different factors (*e.g.* health status and dietary and environmental factors) can influence the potential for an outbreak of mycotoxicosis in animals.[6] The main challenge in undertaking toxicity studies on masked mycotoxins is the limited knowledge regarding the existence of these compounds. Although some of the masked mycotoxins were already recognised a couple of decades ago, most achievements concerning their structural elucidation have been made during the last few years.[28–30] These accomplishments have in part occurred because scientific awareness has led to increased research on masked mycotoxins and partly because the analytical instrumentation, especially mass spectrometry, has strongly evolved. The simultaneous progression of computer software has enabled the identification of a number of masked mycotoxins with high liability. Nonetheless, the chemical identification of a compound is only the first step in the perplexing risk assessment process. Toxicological evaluation of a masked mycotoxin requires purification or (bio)chemical synthesis of the compound in question. As most of these compounds naturally occur in low concentrations, and as many of them have only recently been discovered, very limited amounts of masked mycotoxins are accessible for toxicological evaluation.

It is important to investigate the toxicology and modes of action of mycotoxins using various *in vivo* and *in vitro* models. Masked mycotoxins, for instance, necessitate techniques that enable reliable investigation of their fates within the GIT, their bioavailability and models that take food–toxicant interactions into consideration. In spite of the limited substance availability, research on masked mycotoxins can only benefit from the utilisation of *in vitro* systems for the study of their toxicological properties. To sum up, *in vitro* methods are more suitable than *in vivo* experiments for testing complex test materials, and can be used to investigate individual human differences in susceptibility and polymorphisms in the biotransformation of drugs or toxicants.[31]

6.3.1 Extrapolation to Human Toxicology

An important question is how well *in vitro* data correspond to *in vivo* findings, specifically in association with human health.[32] The linking of *in vitro* toxicity test results with *in vivo* experimental data has been a continuous effort for toxicologists in academia and in regulatory authorities. The dose determined by the concentration of the chemical in the body or at the active sites strongly depends on the rate and extent of chemical absorption and disposition in the body. Two critical elements of chemical absorption are *in vivo* dissolution in the GIT and the transport of chemicals across the intestinal epithelial cells and into the blood stream.[33] While permeation and disposition are inherent properties of a toxicant under normal circumstances, *in vivo* dissolution greatly relies on its chemical properties and route of administration, as well as the biochemical conditions of the GIT.

In vitro methods provide insights at the cellular or molecular level, whereas *in vivo* assays address questions at the systemic level.[34] For example, primary cultures of rat hepatocytes offer several advantages compared to experimental animals when used for chemical biotransformation testing and the evaluation of cellular toxicity mechanisms.[35] For ethical reasons, toxicological experimentation with harmful chemicals in humans is out of the question. Nonetheless, in some cases it is important that inter-species differences in metabolic pathways, the nature of metabolites and relative rates of biotransformation are kept to a minimum, and under these circumstances, *in vitro* assays that utilise human cells or engineered tissues should be favoured.[36] A major advantage of *in vitro* systems in which human cells and tissues are used is that they eliminate the need for extrapolating data from laboratory animals to humans. Consequently, results obtained from the toxicity testing of masked mycotoxins with human tissues are directly applicable in human risk assessment.

One needs to keep in mind that *in vitro* methods, besides their many advantages, have certain limitations in direct extrapolation to *in vivo* toxicology. *In vitro* assays typically disrupt cellular structural integrity and intercellular communication after dissociation of the tissue. Additionally, *in vitro* and *in vivo* toxicokinetics are different, and difficulties will arise in the estimation of the toxicologically active and relevant *in vivo* dose. *In vitro* assays typically increase our understanding of the mechanisms involved in chemical-induced toxicity, because *in vivo* models are complicated by the presence of structural and functional heterogeneity, which does not allow independent examination and interpretation. Of course, if it is known that a particular chemical is selectively toxic to a target organ, one should use that tissue in an *in vitro* test system. *In vitro* studies should be conducted with dose levels and exposure times similar to those in *in vivo* experiments so that dose–response and time–response effects are properly demonstrated.

Several additional factors should be considered when interpreting data from *in vitro* experiments in terms of *in vivo* findings. For example, chemicals in *in vitro* assays are frequently administered in buffers and/or media that strongly differ from the conditions *in vivo*, especially in experiments relating concentration to effect. Toxicologists typically use the area under the curve (AUC) as the key associative or causative parameter for toxicity in experimental animal studies. In general, this is appropriate if target-organ toxicity studies are performed, such as hepatocellular toxicity. However, when investigating other toxicity endpoints such as inhibition, which is regarded as a threshold event, the peak chemical concentration is more relevant. False-negative results or data suggesting a lack of interaction between a chemical and a pre-defined endpoint must be interpreted with caution. Chemicals that are particularly insoluble in aqueous solutions and are used for *in vitro* assessments often indicate low toxicity during *in vitro* screening, because the dose is actually lower than the calculated chemical concentration in the solution.

As a first approach, metabolic conversions of xenobiotics are examined using subcellular fractions. These systems usually only favour the specific biotransformation step, depending on the type of isolation procedure, the cofactors added, the source of the tissue and the expression level of the enzymes involved. However, the balance between metabolic activation and inactivation requires a highly ordered interplay of many enzymes and cofactors in most cases. Currently, the common practice to overcome these limitations is the use of recombinant human enzymes with the necessary metabolic capacity. Use of more integrated systems, such as cells or tissues, can be an alternative solution. One of the major difficulties is determining the target cell concentration, due to the absence of effective toxicokinetic data. Another challenge to overcome is how to mimic neural, immune and endocrine systems in order to be able to study the effects in these organ systems *in vitro*.[37] Commonly used cell lines in *in vitro* assays are usually transformed; in other words, the cells are derived from tumours. Immortalised cells can be kept in culture media from a few hours to a few days. In many cases, they will partially or permanently lose their differentiated properties while in culture media. These intrinsic weaknesses result from the fact that, *in vitro*, the cells are isolated from their natural environment and are no longer integrated into an ordered tissue and organ topology. This results in reduced survival, an imbalance in xenobiotic metabolism and other side effects that can interfere with the performance of an *in vitro* assay.

The general notion of using *in vitro* tests to predict *in vivo* effects should not, however, be limited to the most common applications of *in vitro–in vivo* correlations. Permeation studies with various *in vitro* epithelial cell cultures, using cultured human tissues or excised animal intestine membranes, or even synthetic membranes, have increased our understanding of the permeation properties of many chemicals.[33] Generally, *in vivo* responses used in the development of *in vitro–in vivo* correlations ignore inter- and intrasubject variability, which might be an important parameter. Therefore, interpretation of the findings obtained from experimental studies performed on rodent and non-rodent species to assess possible risks to human health is complicated by inter-species differences in the GIT.

6.3.2   Cytotoxicity of Masked Mycotoxins

Many mycotoxins (in particular trichothecenes) are inhibitors of protein synthesis,[9,38,39] and rapidly dividing cells and tissues with a high protein turnover may consequently become the main targets of these toxicants or their derivatives. Intestinal cells are constantly dividing and are apparently the first to be exposed to mycotoxins after the ingestion of contaminated food or feed, often at higher concentrations than most other tissues. Toxic effects may therefore be caused by mycotoxins to the intestinal epithelium either before absorption in the upper GIT or throughout the entire intestine as a result of non-absorbed mycotoxins. The same is true for masked mycotoxins, with the addition that these compounds may be reactivated to their parent forms, enzymatically or by the action of intestinal microbiota within the lumen. However, for most toxicological assays (genotoxicity, short-term and long-term toxicity, including carcinogenicity, reproductive and developmental toxicities), no studies have so far been conducted for this group of compounds.[40] Due to the fact that the active sites of toxic groups most often remain intact after conjugation processes, the inherent toxicity of masked mycotoxins must be thoroughly investigated.

In the case of deoxynivalenol-3-glucoside (DON-3-Glc), for instance, the epoxide group that is critical for its toxic properties is not affected when a glucose moiety is attached to DON during *in planta* metabolism. However, glucosidation or glucuronidation reactions are mainly considered as detoxification processes in plants and animals. Therefore, it comes as no surprise that DON-3-Glc was shown to have a strongly reduced ability to inhibit protein synthesis by ribosomes, in a wheat germ extract-coupled transcription/translation *in vitro* system, in comparison to DON.[41] Findings from several studies indicate that there is a strong correlation between the *in vitro* resistance of wheat cultivars to DON and *Fusarium* Head Blight (FHB) resistance in the field.[42,43] In FHB-resistant wheat lines, the applied DON is effectively converted to DON-3-Glc as detoxification product, revealing a close relation between the DON-3-Glc/DON ratio and DON resistance in wheat ears.[44] In addition to being a plant metabolite, DON-3-Glc might also be produced by certain fungi from DON.[45] Only indirect evidence has been provided for that hypothesis though.

According to our recent unpublished *in vitro* cytotoxicity study on human Hep-G2, Caco-2 and rat hepatoma H-4-II-E cell lines, the toxicity of DON-3-Glc was also significantly lower than that of its parent toxin, which caused inhibition of cellular metabolic activity by up to 47% (unpublished data). Cytotoxicity was determined with the colorimetric microplate-based alamarBlue® assay after treatment of cells for up to 48 h with DON (1.5–9.0 µM) and DON-3-Glc (3.0-21.0 µM). Hence, alterations in the ability of the conjugated mycotoxin to enter the cells and bind to ribosomes or receptors may explain the difference in cell toxicity and toxicokinetics observed in DON-3-Glc compared to the native toxin.[12] Some other mycotoxin conjugates of DON can also be directly excreted by fungi, such as 3-acetyl-DON (3-Ac-DON) and 15-acetyl-DON (15-Ac-DON).[46] Due to their toxic potential, the Joint FAO/WHO Expert Committee on Food Additives (JECFA) decided to include the acetylated derivatives 3-Ac-DON and 15-Ac-DON in the provisional maximum tolerable daily intake (PMTDI) of 1 µg kg^{-1} b.w. DON, as additional contributing factors to dietary exposure to the mycotoxin.[47] In this evaluation, JECFA experts pointed out the lack of toxicological data for DON-3-Glc, as well as for the other masked mycotoxins and suggested that studies on absorption, distribution, metabolism and excretion (ADME) are needed.

Masked forms of ZEN are the only other group of derivatives, besides those of DON, that have to some extent been studied for their toxicity. ZEN is commonly found in maize, but also in barley, oats, wheat, rye, sorghum and rice, and has a strong oestrogenic activity in vertebrates.[48] It is rapidly and extensively absorbed from the mammalian GIT and exerts its hormonal effect by binding to the oestrogen receptors ER-α and ER-β.[49] It is evident from the literature that ZEN has a relatively low acute toxicity following oral administration, with LD_{50} values of more than 2000 mg/kg b.w. in mice, rats and guinea pigs. Subchronic and long-term toxicity experiments have confirmed its oestrogenic effects with indications of hepatic disturbances and haematological changes in rodents, as well as oestrogenic activity in humans.[22] *In vitro* assays have revealed that ZEN exerts immunotoxic effects by inhibiting rat and human peripheral blood lymphocyte proliferation.[50,51] JECFA has proposed a PMTDI value of 0.5 $\mu g\,kg^{-1}$ b.w. for ZEN.[52]

Detoxification pathways of ZEN result in significant amounts of masked forms of this mycotoxin in *Fusarium*-infected plants. It has been shown that ZEN is transformed to at least 17 different metabolites in *Arabidopsis thaliana* producing among others α-zearalenol (αZEL), an even more potent oestrogenic compound than the parent, and β-zearalenol (βZEL), a metabolite with a lower affinity for oestrogen receptors in comparison to ZEN.[53] In addition to its major metabolites, the zearalenols, other plant-specific metabolites are also present such as zearalenone-14-glucoside (ZEN-14-Glc), α-ZEL-14-glucoside (αZEL-14-Glc), β-ZEL-14-glucoside (βZEL-14-Glc), ZEN-14-sulfate (ZEN-14-S), malonyl-glucosides, di-hexose and hexose-pentose disaccharides of ZEN.[54] A recently discovered masked form of ZEN is zearalenone-16-glucoside ZEN-16-Glc.[55] Fungal glucosylation of ZEN has also been described in the literature.[56,57] Poppenberger *et al.* (2006)[58] tested the ability of ZEN-14-Glc to bind to human oestrogen receptors with an *in vitro* competitive binding assay and compared it to that of the precursor mycotoxin. According to the results, attachment of the glucose moiety prevented interaction of the mycotoxin with human oestrogen receptors, yielding far lower oestrogenic activity than that of ZEN. Similar reduced interaction with human oestrogen receptors was observed for ZEN-14-S *in vitro*,[18] although the same compound exerted oestrogenic activity in a rat-feeding test.[59] This controversy might be explained by the *in vivo* enzymatic deconjugation of, for instance, sulfatases that release the native toxin responsible for toxicity.

Fumonisins are structurally related hepatotoxic and nephrotoxic *Fusarium* (and *Aspergillus*) mycotoxins. Fumonisin B_1 (FB_1) is commonly found in maize, rice, sorghum and soybeans, and can cause diseases in animals (leucoencephalomalacia in horses and porcine pulmonary oedema).[60] Fumonisins have been found to interact with the food matrix, forming bonds that make them undetectable by routine analytical methods. These forms include covalently bound derivatives referred to as simply 'bound fumonisins' or non-covalently bound, the 'hidden fumonisins'.[61] It was previously believed that fumonisin conjugates were only formed by interactions with sugars, amino acids and proteins during food processing, however, they can also be found in unprocessed maize.[62] The masking mechanism has been attributed to the formation of covalent bonds between tricarballylic groups of fumonisins and the hydroxyl of starch or the amino groups of the side chains of amino acids.[1] These bound, or hidden, fumonisins can be released by enzymatic hydrolysis during digestion, resulting in higher exposure levels than estimated.[63] An additional implication is that fumonisin derivatives have an increased oral bioavailability and might be converted back to FB_1, or another toxic metabolite, after entering the systemic circulation.[64]

Finally, ochratoxin A (OTA) is a mycotoxin with a controversial toxicological status in the sense that toxic effects associated with its abundance cannot be explained on the basis of its known biochemical mode of action. It is thus assumed that OTA derivatives may have a synergistic contribution to the overall toxicity of contaminated commodities.[1] OTA is found to transform into ochratoxin α (OTα), OTA-methyl-ester and two isomers of hydroxy-ochratoxin A, as well as the β-glucosides and methyl esters of both of these isomers in wheat and maize cell suspension cultures.[65] OTα is regarded as non-toxic, in contrast to hydroxy-ochratoxin A, which has immuno-suppressive effects similar to its precursor; toxicity of the other derivatives remains unknown.[40] Other masked mycotoxins are constantly being discovered, but no information is available regarding their toxicity. The toxic significance of masked mycotoxins may not only relate to their inherent toxicity, as hydrolysis to the parent compound could be the main concern. However, even if they are hypothetically of lower toxicity, the masked forms of mycotoxins need to be investigated in more detail to determine their bioavailability and toxicological relevance before any final conclusions are drawn.

6.4 Bioaccessibility and Bioavailability Assessment *In Vitro*

Digestion is a physiological process starting in the mouth, following the ingestion of food or liquids, and continues until absorption or elimination of the ingested substances. The breakdown of foodstuffs occurs within the GIT, which extends from the mouth to the anus, also including most of the pharynx, oesophagus, stomach, small intestine and large intestine. After ingestion, and as food enters the stomach it is mixed with secretions of the gastric glands, forming a soupy liquid, *i.e.*, chyme that mainly consists of partially digested food, water, digestive enzymes and hydrochloric acid. Thereafter, chyme slowly passes through the pyloric sphincter into the small intestine, where it is further mixed with intestinal juices, as well as pancreatic and bile secretions that aid in the absorption of compounds by microvilli when chyme comes into contact with them.[66] The small intestine is the longest section of the digestive tract and possesses a vast surface area for the absorption and possible metabolism of chemical substances. Enzymatic breakdown and absorption of substances are also assisted by peristalsis along the GIT.

Human and animal exposure to mycotoxins and their derivatives mainly occurs by the oral route, as earlier discussed, making food and feed major sources of health-related risk for these compounds. Nevertheless, the total intake of ingested contaminants does not necessarily reflect the actual amount reaching the systemic circulation by being readily (bio)available to the body. Only the fraction of a contaminant that is released from the food matrix during digestion and becomes (bio)accessible to intestinal absorption may exert toxic effects. More specifically, the oral bioavailability of a (masked) mycotoxin is defined as the percentage of the ingested compound that is released from the food matrix, crosses the intestinal epithelial cells and reaches the systemic circulation *via* the liver.[67] When a xenobiotic is absorbed, it is transported to the liver prior to entering circulation in a process referred to as the first-pass effect. It is assumed, according to this definition, that potential toxic effects are induced by the parent compound and not by metabolic products formed after ingestion. Oomen *et al.*[68] conceptualised the oral bioavailability (F) of mycotoxins as the resultant of three major processes, represented as fractions of: bioaccessibility (F_B), transport across the intestinal epithelium (F_A) and the first-pass effect (F_H). This can be expressed with the following formula:

$$F = F_B \times F_A \times F_H \tag{6.1}$$

After ingestion of food, the contaminants present may be partially or totally liberated from the matrix to the GIT. The matrix of ingestion may lower the bioaccessible fraction ($F_B < 1$) and thus lower internal exposure. The proportion of any contaminant that is mobilised from the food into the digestive juice is defined as bioaccessibility (B), and represents the maximum amount of a contaminant available for transport from the lumen across the intestinal epithelium and into the portal vein or lymph. This concept is only applicable to oral exposure and is estimated in percentages with the equation:[69]

$$B(\%) = \frac{[\text{mycotoxin}_{\text{chyme}}] \text{ after GI digestion}}{[\text{mycotoxin}_{\text{food matrix}}] \text{ before GI digestion}} \tag{6.2}$$

It is worth mentioning that bioaccessibility can only be estimated with *in vitro* systems, because *in vivo* determination would require samples to be taken from the small intestine at different sites and time points, which is generally not achievable.[67] The matrix and concentration level in which a toxicant is present plays an important role in bioaccessibility, whereas absorption and metabolism are more compound-specific properties and dependent on mammalian physiology.[70] Bioaccessibility of mycotoxins has been demonstrated to differ according to the considered food matrices, *e.g.* the bioaccessibility of DON differs between various types of commercial pasta.[71] The bioavailability of mycotoxins can also vary from the maximal level, as with aflatoxins, to a very limited level, as with fumonisins, and differs between mammalian species.[72] Therefore, knowledge of bioaccessibility and bioavailability of an ingested contaminant can in some cases be more relevant than the total concentration present in a food sample and are both essential in the toxicological evaluation of any given compound for human health risk assessment.

In vivo methods provide direct information on the bioavailability of a compound by measuring changes in its concentration in the blood plasma as a function of time (*e.g.* AUC), following oral exposure. Drawbacks of *in vivo* bioavailability methods include ethical restrictions, increased costs and time, and possibly high biological variability among test subjects. *In vitro* gastrointestinal methods are widely used alternatives to *in vivo* assays for bioaccessibility/bioavailability assessment capable of simulating the digestion and absorption processes under laboratory conditions.[73] Gastrointestinal models are safe, avoid the ethical issues and excessive documentation associated with *in vivo* studies and can be performed in a rather short period of time. Despite the general progress in *in vitro* methodology, challenges still exist. One of the main limitations of *in vitro* digestive assays lies in the fact that the mammalian digestive system and brain are equipped with a series of inherent barriers and defence mechanisms against non-essential compounds, toxicants and microorganisms, which *in vitro* models are currently unable to simulate.[66]

Several models have been utilised to study the release and absorption of a number of compounds (*e.g.* carbohydrates, proteins, antioxidants and phenolic compounds).[74] Two types of *in vitro* gastrointestinal models have been used in the study of masked mycotoxins: assays for the determination of bioaccessibility only by simulating the digestion process, and assays for the determination of bioavailability by also integrating the absorption process using Caco-2 cell cultures. Both types of models are capable of investigating the partial or complete hydrolytic fate of masked mycotoxins, depending on the compartments used. Analyses of the samples obtained in digestive studies are commonly performed with analytical techniques based on liquid or gas chromatography coupled with mass spectrometry. Figure 6.2 illustrates a schematic representation of an *in vitro* gastrointestinal model for the study of bioaccessibility and bioavailability of masked mycotoxins.

6.4.1 Human Digestion Models

To determine bioaccessibility, *in vitro* digestion models have been developed that are able to mimic, in a simplified manner, the human GIT and its physiological functions. Human digestion models have proven to be valuable experimental tools in assessing the potential risk from ingested xenobiotics. The majority of *in vitro* digestion models that have been used for assessing the bioaccessibility of masked mycotoxins have been performed based on the protocol of Versantvoort *et al.* known as the RIVM model.[75] The chemical composition of digestive fluids, pH values, temperature and transient times are reproduced, simulating the actual conditions during the most important gastrointestinal digestion steps. The main parameters to

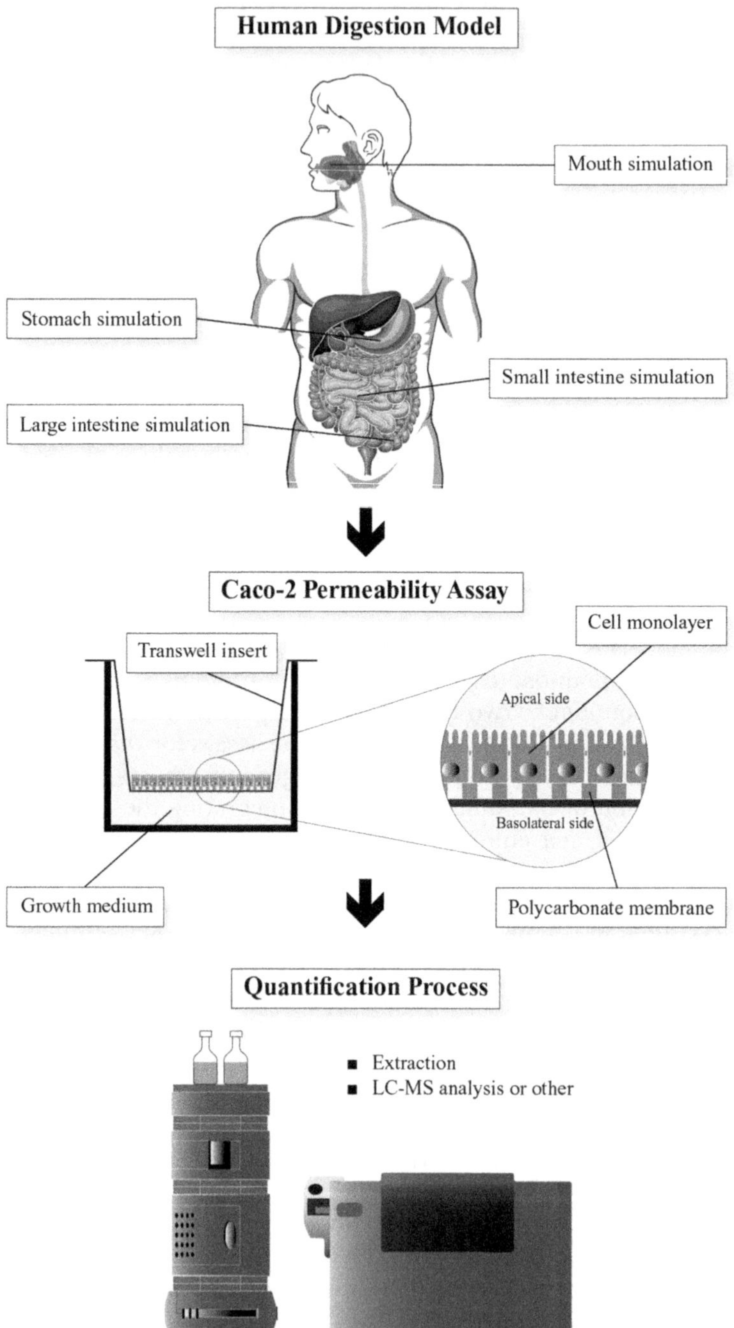

Figure 6.2   *In vitro* gastrointestinal model setup to determine the bioavailability of masked mycotoxins, involving human digestion and Caco-2 permeability assays. Quantification performed with liquid chromatography-mass spectrometry (LC-MS).

control are temperature and digestive juice composition, as well as time and the pH of each compartment. In the human digestion models, all synthetic digestive juices have been kept at $37 \pm 2\,°C$, as temperature is crucial to enzyme activity and chemical characteristics (*e.g.* analyte solubility). The volumes of the various digestive juices have been based on physiology, resulting in a ratio of $1:1.5:3:1$ for saliva, gastric juice, duodenal juice and bile, respectively.[76] Systematic mixing of the matrix with the digestive juices is also important. The constituents of the digestive juices and processes are described in the following sections (Table 6.1).

6.4.1.1   Mouth and Stomach Simulation

Mechanical digestion in the mouth results from mastication, whereas the chemical breakdown of food is initiated within the oral cavity by the secretion of saliva from the salivary glands. The production of saliva and its flow rate are increased when food is present in the mouth, by certain visual and/ or olfactory stimuli and chewing. In general, human saliva consists of approximately 99.5% water, while other solutes such as electrolytes (sodium, potassium, chloride, calcium, bicarbonate, magnesium and phosphate), enzymes (amylases and lipases), immunoglobulin A and other antimicrobial factors account for the remaining 0.5%. Some dissolved gases and various other organic substances, including urea, uric acid and mucin are also present.[77] The water in saliva provides a medium for dissolving food components so that digestion reactions can begin. Salivary amylase is activated by the presence of chloride ions in saliva to begin the hydrolysis of α-(1,4)-linked polysaccharides. Lingual lipase is responsible for initiating the hydrolysis of dietary lipids to the corresponding diglycerols. Artificial saliva used in several of the digestion models for masked mycotoxins is usually prepared as an aqueous solution of electrolytes (KCl, KSCN, NaH_2PO_4, $NaHCO_3$, NaCl and $NaHCO_3$), α-amylase, urea, uric acid and mucin. The pH of saliva in adults typically ranges from 6.0 to 7.0,[78] but in digestion experiments the pH has been adjusted to 6.8 ± 0.2. Contaminated food, standard meals or ground cereals have been used as test matrices. Depending on the amount of sample, usually between 2 and 4.5 g, 3 to 6 mL of saliva have been added resulting in a sample-to-saliva ratio of $1:2$ (m/v), followed by incubation for 5 min.

Once food reaches the stomach, the gastric phase begins with the secretion of acidic gastric juice and ends when the stomach contents reach the duodenum to start the intestinal phase. In adults, approximately 1.5 L of gastric juice are typically secreted daily.[76] Gastric juice consists of water, enzymes, mineral salts, hydrochloric acid and mucus, and is secreted by the gastric glands in the stomach wall. The secretion of hydrochloric acid lowers the pH of the gastric contents to values between pH 1–3 that is necessary for the activation of pepsin and stimulation of bile flow.[66] Therefore, enzymatic digestion of proteins starts in the stomach by the action of pepsin, which breaks down certain peptide bonds between amino acids, forming smaller

Table 6.1 Constituents and processes involved in a human digestion model for *in vitro* bioaccessibility testing of masked mycotoxins.[87]

	Mouth simulation	Stomach simulation	Small intestine simulation		Large intestine simulation	
	Saliva (3 mL) + Sample (2 g)	Gastric juice (6 mL)	Duodenal juice (6 mL)	Bile juice (3 mL)	Growth medium (2 mL)	Faecal slurry (2 mL)
Constituents	5 mL KCl (89.6 g L^{-1}) 5 mL KSCN (20 g L^{-1}) 5 mL NaH$_2$PO$_4$ (88.8 g L^{-1}) 5 mL Na$_2$SO$_4$ (57 g L^{-1}) 850 μL NaCl (175.3 g L^{-1}) 10 mL NaHCO$_3$ (84.7 g L^{-1}) 4 mL urea (25 g L^{-1}) 290 mg α-amylase 15 mg L^{-1} uric acid 25 mg L^{-1} mucin	7.85 mL NaCl (175.3 g L^{-1}) 1.5 mL NaH$_2$PO$_4$ (88.8 g L^{-1}) 4.6 mL KCl (89.6 g L^{-1}) 9 mL CaCl$_2$ (16.65 g L^{-1}) 5 mL NH$_4$Cl (30.6 g L^{-1}) 3.25 mL HCl (37%) 5 mL glucose (65 g L^{-1}) 5 mL glucuronic acid (2 g L^{-1}) 1.7 mL urea (25 g L^{-1}) 5 mL glucosamine hydrochloride (33 g L^{-1}) 1 g L^{-1} BSA 2.5 g L^{-1} pepsin 3 g L^{-1} mucin	20 mL NaCl (175.3 g L^{-1}) 20 mL NaHCO$_3$ (84.7 g L^{-1}) 5 mL KH$_2$PO$_4$ (8 g L^{-1}) 3.15 mL KCl (89.6 g L^{-1}) 5 mL MgCl$_2$ (5 g L^{-1}) 90 μL HCl (37%) 2 mL urea (25 g L^{-1}) 9 mL L^{-1} CaCl$_2$ (16.65 g L^{-1}) 1 g L^{-1} BSA 9 g L^{-1} pancreatin 1.5 g L^{-1} lipase	15 mL NaCl (175.3 g L^{-1}) 34.15 mL NaHCO$_3$ (84.7 g L^{-1}) 2.1 mL KCl (89.6 g L^{-1}) 75 μL HCl (37%) 5 mL urea (25 g L^{-1}) 10 mL L^{-1} CaCl$_2$ (16.6 g L^{-1}) 1.8 g L^{-1} BSA 30 g L^{-1} bile	5 g soluble starch 5 g peptone 5 g tryptone 4.5 g yeast extract 4.5 g NaCl 4.5 g KCl 2 g pectin 4 g mucin 3 g casein 2 g arabinogalactan 1.5 g NaHCO$_3$ 0.69 g Mg$_2$SO$_4 \cdot$ H$_2$O 1 g guar 0.8 g L-cysteine HCl$\cdot$H$_2$O 0.5 g KH$_2$PO$_4$ 0.5 g K$_2$HPO$_4$ 0.4 g bile salt 0.08 g CaCl$_2$ 0.005 g FeSO$_4 \cdot$ 7H$_2$O 1 mL tween 80 Resazurin solution (0.025% w/v)	Fresh faecal samples Dulbecco's phosphate buffer saline (10% v/v) Toxin (5 mg L^{-1})
pH	6.8 ± 0.2	1.30 ± 0.02	8.1 ± 0.2	8.2 ± 0.2	–	
Temperature	37 ± 2 °C	37 ± 2 °C	37 ± 2 °C	37 ± 2 °C	37 °C	
Incubation	5 min	2 h	2 h		24 h	
Mixing	250 rpm magnetic stirrer	250 rpm magnetic stirrer	250 rpm magnetic stirrer		200 strokes per min in Dubnoff bath	
Other	–	–	1 M Bicarbonate solution (1 mL)		Anaerobic conditions	

peptides.[79] Mucin, a large glycoprotein, is another key component during digestion due to its ability to form gels that protect and lubricate the GIT. During gastric emptying, and as the chyme passes into the duodenum, pepsin is denatured and the gastric pH gradually declines until the fasted state (pH 1.5–2) has been re-established. Gastric emptying is determined by the volume, osmotic pressure and caloric content of the meal. In most *in vitro* models used for masked mycotoxin bioaccessibility assessment, gastric juice is simulated as a mixture of several salts (NaCl, NaH_2PO_4, KCl, $CaCl_2$, NH_4Cl and HCl), glucose, glucuronic acid, urea, glucosamine hydrochloride, bovine serum albumin (BSA), pepsin and mucin. Gastric pH has been kept constant and low with values ranging between pH 1.30 ± 0.02 and 2.5 ± 0.5. Gastric pH is a crucial factor in the determination of bioaccessibility, because it is essential for the activity of pepsin, which can contribute to the potential release of masked mycotoxins from the food matrix.[61]

6.4.1.2   *Intestinal Simulation*

Following gastric digestion, the food is transferred past the pyloric sphincter into the duodenum, which is the first compartment of the small intestine. The rest of the small intestine, located below the duodenum, consists of the jejunum and the ileum. Gastric emptying into the duodenum is determined by the type and volume of food ingested. The duodenum receives pancreatic enzymes from the pancreas and bile from the liver and gallbladder. The small intestine is highly effective at absorbing elements essential to the organism, with 80% of the chyme being absorbed before entering the colon. Its high absorption efficiency makes it a target of entry for bacteria, xenobiotics and toxic substances. Each day, hepatocytes in the liver secrete 400–800 mL of bile, a yellowish liquid consisting of water, bile acids and salts, cholesterol and several ions.[80] Bile can have a huge impact on bioaccessibility, because it plays a role in emulsification, the breakdown of large lipid globules, and also lowers surface tension, enhancing the solubility of hydrophobic compounds.[69]

The components of intestinal juices are extremely complex to reproduce, and *in vitro* models consequently employ simplified versions of them. Mixtures mimicking duodenal (6–12 mL) and bile (3–6 mL) juices have been added to the mixture at this stage. Duodenal juice basically consists of the main duodenal electrolytes (NaCl, $NaHCO_3$, KH_2PO_4, KCl, $MgCl_2$, HCl and $CaCl_2$), urea and BSA as well as the enzymes pancreatin and lipase. Bile juice is a mixture of bile salts (NaCl, $NaHCO_3$, KCl, HCl and $CaCl_2$), urea, BSA and bile. Together with the two intestinal juices, 1–2 mL of bicarbonate solution (1 M) is added to the mixture and a final 2-h incubation step is performed. The intestinal secretions are produced by enterocytes and have a slightly alkaline pH in the range of 7.5 to 8.0.[76] The *in vitro* models discussed here, employed a pH of 8.1 ± 0.2 and 8.2 ± 0.2 for duodenal and bile juices, respectively.

6.4.1.3 Gastrointestinal Microbiota

The GIT of humans and animals contains vast quantities of microorganisms, existing in symbiosis with the host, with a virtually unlimited metabolic potential. The intestinal microbiota is a complex ecosystem that is unique for each organism, fulfilling a variety of vital physiological functions that have a major contribution to an individual's health and well-being. Bacteria make up most of the microflora, with estimates of more than 500 species inhabiting the human gut.[81] The digestion and absorption of nutrients such as carbohydrates, proteins, amino acids, peptides, fats and vitamins is highly influenced by the presence of a healthy microbial flora. These microorganisms contain enzymes also capable of hydrolysing undigested compounds that the upper GIT of humans is unable to process. Additionally, the gut microbiota plays an important role in the immune system, prevents the colonisation of pathogens and contributes to the break-down, metabolism and transformation of xenobiotics. It is known, that the human or animal colon microbiota is able to break down certain mycotoxins (*e.g.* OTA, AFB_1, DON, *etc.*).[82]

The localisation and size of bacterial populations varies considerably between animal species. In polygastric animals (*e.g.* cattle and sheep), a high bacterial content is located both before and after the small intestine, whereas in most of monogastric species, including humans, pigs, dogs and rodents, the microbial flora only exists after the small intestine, in the colon. It is therefore obvious that species differences in bacterial populations can create major differences in the digestive fate of xenobiotics between monogastric and polygastric species. As an example, the presence of large numbers of bacteria that are able to convert toxic DON into its non-toxic de-epoxide metabolite, DOM-1, before the small intestine in ruminants and poultry, massively reduces the amount of native DON reaching the small intestine, making such animal species almost insensitive to oral intoxication by DON.[12]

Xenobiotics, and especially antibiotics, can affect the total count of bacteria in the gut and also the relative populations of microbes. Toxicants can additionally affect the microflora by eradicating selective or broad populations within the lumen of the GIT. This could lead to a loss of homeostasis or total obliteration of intestinal bacteria in some extreme cases.[66] Resident bacterial populations appear to be relatively stable to changes in the diet. Gut flora can also metabolise xenobiotics, forming toxicologically inactive or active metabolites with unknown consequences. The activation of toxic metabolites by mycotoxin metabolism or de-conjugation of masked mycotoxins by intestinal microbiota may lead to the formation of compounds that undergo enterohepatic circulation, which would subsequently result in an increased exposure. The interaction of digestive microbiota with parent mycotoxins has been predominantly studied in models simulating the digestive tract of ruminants and focusing on the degradation of these toxins.[69] Studies regarding interactions between intestinal flora and masked mycotoxins have been either performed using pure cultures of isolated bacterial strains or with anaerobic faecal fermentation utilising the human colon microbiome. As we discuss in the following section, it is very important not to neglect microbial interactions when investigating the bioaccessibility of masked mycotoxins with *in vitro* models.

6.4.1.4 Digestive Fate of Masked Mycotoxins

In general, mycotoxins can be divided into three groups based on their bioaccessbility profile. Specifically, aflatoxins, enniatins and fumonisins (despite their low bioavailability) belong to the group of mycotoxins with high bioaccessibility, with values ranging between 70–100%, DON and patulin have intermediate values (30–70%), whereas ZEN usually presents low bioaccessibility. OTA shows high variability on its bioaccessibility, depending on the matrix (30–100%).[69] Knowledge of bioaccessibility, as previously mentioned, is an important parameter in risk assessment. However, abundance of masked mycotoxins in test samples may lead to interferences in the estimation of native mycotoxin bioaccessibility, a fact that should be taken into account. Findings from an *in vitro* study, utilising the RIVM model, reported similar recoveries from the chyme for DON and DON-3-Glc with values of 65% and 55%, respectively.[83]

As the general toxicity of the known masked mycotoxins is most presumably low, their reactivation during the mammalian digestion should give rise to further consideration. In 2011, JECFA acknowledged the possibility that DON-3-Glc may be hydrolysed in the mammalian GIT, increasing exposure to the precursor mycotoxin.[47] Therefore, exposure to certain mycotoxins may be underestimated if masked mycotoxins are reactivated within the GIT, for instance enzymatically. It is well known that salivary amylase is able to cleave the α-glucosidic bonds of starch, and that a variety of compounds secreted in the bile, *e.g.* glucuronides or glucosides, can be hydrolysed by the action of enzymes such as β-glucuronidase and β-glucosidase.[84] However, based on *in vitro* experimental data obtained from human digestion models, acidic or enzymatic hydrolysis of DON-3-Glc does not occur under circumstances found in the stomach or upper GIT of mammals.[83,85,86] The de-conjugation of DON-3-Glc after the small intestine in monogastric animals minimises the possibility of absorption of the released toxin, as the uptake of DON mainly takes place in the duodenum. Masked forms of ZEN (ZEN-14-Glc and ZEN-14-S)[87] and covalent fumonisin conjugates (*N*-alkyl and acyl conjugates),[88] have also have proven to be stable under the conditions found in the upper GIT. On the other hand, fumonisins non-covalently bound to proteins and carbohydrates of the food matrix are a special type of masked mycotoxins.[63] These compounds can be released early in the digestive process, making the parent compounds already available for absorption and/or toxicity in the upper GIT. However, no specific bioavailability studies have thus far been conducted. An interesting observation was made for HT-2 toxin-3-glucoside (HT2-3-Glc), while performing *in vitro* digestion of T-2 toxin (T2) and HT-2 toxin (HT2), as it is the first indication that a masked mycotoxin is released during simulation of the gastric phase in a gastrointestinal model (IFR assay).[89] Unfortunately, there are no quantitative data in this study and observations are based on relative peak areas. Table 6.2 summarises all findings obtained by different *in vitro* digestion models concerning the release of masked mycotoxins within the GIT.

During the past few years, a number of studies have highlighted the significance of the intestinal flora in the chemical modification of masked mycotoxins. Initially, the partial conversion of DON-3-Glc to DON was reported after incubation with strains of intestinal bacteria and particularly species of the genera *Lactobacillus*, *Enterococcus*, *Enterobacter* and *Bifidobacterium*.[86] Although hydrolysis of DON-3-Glc at a late phase of the digestion process might be of minor toxicological relevance, late absorption of other de-conjugated masked mycotoxins cannot be excluded. Findings from a faecal fermentation assay demonstrated rapid (within 30 min) and complete de-conjugation of ZEN-14-Glc and ZEN-14-S by human colonic microbiota.[87] In the study by Dall'Erta *et al.*,[87] an identical hydrolytic fate for DON-3-Glc was observed, although not as rapid as in the case of masked ZEN forms, but within the physiological residence time in the human colon. These findings for DON-3-Glc de-conjugation are in line with the conclusions from a similar study also utilising human colonic microbiota.[90] In this work, it was also reported that only a small minority of the test group were able to degrade free DON to its less toxic metabolite DOM-1, starting after 6 h of incubation by bacterial action. The lack of human colonic microbial detoxification stresses the fact that the human intestine is unprotected against the toxic effects of DON. The toxicological significance of this exposure warrants further investigation to better assess the role of masked DON in human and animal toxicology, as it might induce certain neuroendocrine effects.[11]

ZEN-16-Glc was also rapidly and entirely converted to ZEN due to microbial action following human faecal fermentation. No remaining ZEN-16-Glc was found in the spiked faecal slurry after 2 h incubation. The authors argued that either periplasmatic β-glucosidases or cytosolic glucoside hydrolase family members of the bacteria are the intestinal hydrolytic enzymes responsible for cleavage of the aglycone.[55] Based on *in vitro* and *in vivo* data for the digestive fate of ZEN-14-Glc and ZEN-16-Glc, it appears likely that the oestrogenic effects of these masked forms are equally toxicologically relevant to those of ZEN. Possible intestinal reactivation of masked mycotoxins not only poses a risk to the primary target organ of the parent compound, but the tissue at the site of de-conjugation may also be affected. Late-phase hydrolysis of the masked mycotoxins DON-3-Glc, ZEN-14-Glc, ZEN-16-Glc and ZEN-14-S by bacteria in the large intestine and the consequent release of the aglycones expose the colonic epithelium to significant amounts of free mycotoxins. Additionally, hidden fumonisins might be a substrate for bacterial action and could be hydrolysed to give FB_1, and (partly)HFB_1.[91] The toxicological implications may be entirely different in

Table 6.2 Release of masked mycotoxins by acidic, enzymatic or microbial hydrolysis, determined by *in vitro* human digestion models.[a]

In vitro digestion model	Matrix	Masked mycotoxin(s)	Maximum acidic-enzymatic cleavage (w/w)	Maximum microbial cleavage (w/w)	Ref.
Enzymatic treatment	No matrix	DON-3-Glc	<LOQ	Not tested	91
RIVM	Corn flakes	Hidden fumonisins	37–64% increase in total fumonisin content after 4 h	Not tested	98
RIVM	Raw maize	Hidden fumonisins	30–50% increase in total fumonisin content after 4 h	Not tested	69
Various independent assays	No matrix	DON-3-Glc	<0.2%	62% after ≤8 h	92
RIVM	Maize products	Hidden fumonisins	70-99% increase in total fumonisin content after 4 h	Not tested	94
RIVM	Infant formula	DON-3-Glc	<5.0% after 4 h	Not tested	89
RIVM + human faecal fermentation	Ground sample	DON-3-Glc	≤0.5% after 4 h	≈100% after ≤24 h	93
		ZEN-14-Glc	≤2.7% after 4 h	≈100% after ≤30 min	
		ZEN-14-S	≤1.4% after 4 h	≈100% after ≤30 min	
Human faecal fermentation	No matrix	DON-3-Glc	Not tested	≈100% after ≤6 h	96
IFR	Bread	HT2-3-Glc	Almost complete after 120 min[b]	Not tested	95
Human faecal fermentation	No matrix	ZEN-16-G	Not tested	≈100% after ≤30 min	97

[a]Abbreviations: DON-3-Glc, deoxynivalenol-3-glucoside; LOQ, limit of quantification; ZEN-14-Glc, zearalenone-14-glucoside; ZEN-14-S, zearalenone-14-sulfate; HT2-3-Glc, HT-2 toxin-3-glucoside; ZEN-16-G, zearalenone-16-glucoside.
[b]Semi-quantitative data.

ruminants and poultry, in which, de-conjugation may take place before the small intestine, potentially allowing the absorption of the released toxin. Similar biotransformations are expected to yield even more severe effects in the case of other mycotoxins, such as T2, which is one of the agents that are destructive to mucosal surfaces, causing the disruption of secretory and digestive functions.[92] Nonetheless, based on the current knowledge as regards digestive stability of masked mycotoxins, DON-3-Glc seems to be more stable than ZEN-glucosides and non-covalently bound fumonisins. Research should therefore focus on the identification and mechanisms of formation of mycotoxin conjugates that have a reduced bioavailability, remain stable during digestion and ideally are less toxic compared to the native toxins. Potential candidates that may fulfil such requirements are GSH-based conjugates, which have been reported as stable and less toxic in animals and plants.[93,94]

Finally, interactions such as absorption, metabolism and degradation between mycotoxins and microorganisms located in the gut are well known. However, de-conjugation processes of masked mycotoxins in the lower GIT directly expose the microbial flora to toxic parent compounds, which can disturb the normal population and lead to adverse health effects. Altering the numbers of GIT bacteria may affect the ability of hosts to digest food and to stimulate the immune system.[95] Several mycotoxins elicit toxicity to the gut microbial composition, posing a direct risk to human and animal health. An example of this was provided in a recent study demonstrating significant changes on the composition of the human gut microbial population, after oral subchronic exposure to DON in a model of human microbiota-associated rats.[96] In another experiment, exposure of pigs to T2 resulted in a substantial increase of aerobic bacteria counts in the gut.[97] Therefore, the release of additional amounts of parent mycotoxins in the large intestine may introduce secondary adverse effects as a consequence of combined toxicity between de-conjugated and unabsorbed parent toxins in the gut. Possible increased targeted mycotoxin exposure to microbial flora, due to masked mycotoxins functioning as carriers of toxins, is a completely unexplored area that may be proven as the ultimate threat regarding masked mycotoxins.

6.4.2   Caco-2 Permeability Assay

Human digestion models in combination with other techniques, such as intestinal absorption models with various epithelial cells, increase our understanding on permeation properties of many toxic molecules and offer a more complete picture of the processes undergoing within the GIT. For the past 20 years, the Caco-2 monolayer cell culture model has been extensively used for studying the intestinal uptake of drugs and toxins, as it provides a detailed and mechanistic insight into the metabolism and transport of xenobiotics.[98] Caco-2 cells are a well-differentiated human colon adeno-carcinoma cell line, forming a tight monolayer when grown *e.g.* on Transwell polycarbonate membranes. Although derived from the large intestine, under appropriate culture conditions they differentiate into polarised monolayers, possessing morphological and biochemical characteristics similar to the enterocytes lining the small intestine.[99] These cells express tight junctions, microvilli on the apical side and many small intestine enzymes and transporters (*e.g.* Figure 6.2).

Hidalgo and co-workers[100] first pioneered the application of Caco-2 cells as an *in vitro* model to determine the intestinal permeability of various compounds in humans. For permeability assays, the Caco-2 model is preferred to other colon carcinoma cells because it is performed in a relatively short period of time, in large numbers, under controlled conditions and generates a wealth of information. This model is also extensively utilised for absorption screening purposes and mechanism determination, as well as for metabolic studies. However, it does have some limitations. Caco-2 cells grow slowly compared to other cell lines and can display high variability in expression levels of enzymes and transporter proteins between cell passages.[98] Despite some noted deficiencies, the Caco-2 cell culture model is regarded as a very useful diagnostic tool to determine the permeation/absorption of toxicants in the human intestine without the use of animals or humans. Caco-2 permeability assays allow calculation of the apparent permeability coefficient (Papp) value, which is used to predict the *in vivo* oral bioavailability of chemicals in humans.[99] Extensive studies have demonstrated good correlations between human drug absorption and Papp values determined with the Caco-2 model.[101–103] Therefore, by measuring the permeability of a compound using this assay, the extent of permeation through the intestinal mucosa can be estimated.

6.4.2.1 Transcellular and Paracellular Passive Diffusion

Several mechanism are involved in the transport of chemicals and across cellular membranes, including passive or facilitated diffusion, active transport *via* carriers (ATPases, channel proteins, transporters) or clathrin-dependent and independent endocytosis.[104,105] Toxicants generally traverse membranes by passive diffusion, a mechanism governed by Fick's first law, *i.e.*, chemicals move from regions of higher to regions of lower concentration without energy expenditure. Small hydrophilic molecules (less than 600 Da) permeate the epithelia through aqueous pores within the membrane (paracellular diffusion), whereas larger hydrophobic molecules are absorbed across the lipid domain of membranes (transcellular diffusion). Estimation of absorption has been associated with assessment of the octanol/water partition coefficient, or $\log P$, a very informative physicochemical parameter relative to assessing membrane permeability.[106] It is expressed as the equilibrium ratio of the solute concentrations in the two solvents, octanol and water. Lipophilic compounds that can readily traverse biological membranes have high positive values of P, while negative values suggest a polar behaviour.[3] Another very useful physicochemical property is the distribution coefficient ($\log D$), which reflects the pH-dependent lipophilicity of a chemical. It is defined as the ratio of the un-ionised compound in the organic phase to the concentration of all species in the aqueous phase at a given pH:

$$\log D = \log \left(\frac{[\text{solute}]_{\text{octanol}}}{[\text{solute}]_{\text{water}}^{\text{neutral}} + [\text{solute}]_{\text{water}}^{\text{ionised}}} \right) \qquad (6.3)$$

Both partition and distribution coefficients are measures of how hydrophilic or hydrophobic a chemical substance is, and are useful in estimating the absorption and distribution of drugs or toxins within the body. The diffusion of chemicals through Caco-2 cells has been found to primarily depend on $\log D$ values (when <2.0).[107] Variations in pH along the GIT can markedly alter the permeability characteristics of ionic compounds across biological membranes. The majority of xenobiotics behave as weak acids or bases and are absorbed in the part of the GIT where they exist in their most lipid-soluble form.[3] For example, the ionised form of a molecule usually has low lipid solubility and is thus less likely to be absorbed through the lipid domain of a membrane. In contrast, the un-ionised form of the molecule will have greater absorption, with the rate of transport being proportional to its lipid solubility. However, other factors influence the intestinal absorption of xenobiotics such as the surface area at the site of absorption, the law of mass action, as well as health status of the organism, effects of dietary constituents, food temperature, intestinal motility, rate of emptying and blood flow rate.[66]

6.4.2.2 *Intestinal Absorption of Masked Mycotoxins*

The rapid appearance of mycotoxins in the circulation clearly indicates that the majority of the ingested toxins are absorbed in the proximal part of the GIT.[12,108,109] As previously discussed, the chemical properties of any molecule affect its absorption and therefore masked mycotoxins are expected to behave differently within an organism compared to the parent compounds. The intestinal epithelium is exposed to the entire content of contaminated food or feed and is the first target of these contaminants. The poor intestinal absorption of some mycotoxins implies that the gut epithelium is exposed to a very high proportion of the ingested toxin, especially in the non-ruminant species. Mycotoxins can therefore compromise the intestinal epithelium either before absorption in the upper part or throughout the entire intestine due to the presence of non-absorbed toxins. Most importantly, several mycotoxins and their metabolites undergo enterohepatic circulation making them available again *via* the bile, resulting in reabsorption and a prolonged interaction with epithelial cells.[110–112]

A gastrointestinal model, including the Caco-2 assay in a Transwell system, was used to investigate the possible biotransformation and absorption of DON-3-Glc.[83] One limitation of this setup was that the chyme solution recovered from the *in vitro* digestion model could not be directly applied to the Caco-2 culture because it would damage the cells. Moreover, dilution of the chyme with growth medium would result in concentrations of the analytes below limits of quantification and thus, the Caco-2 cells had to be exposed to DON and DON-3-Glc dissolved in pure growth medium as part of an independent assay. In this experiment, cells were treated with almost equimolar amounts of either DON (3.5 nmol) or DON-3-Glc (3.6 nmol) and samples were collected from both the apical and basolateral sides. After 24 h of exposure, no evidence was found for the hydrolysis of DON-3-Glc to DON, or further degradation to DOM-1 by the Caco-2 cells. Nonetheless, Caco-2 cells are described in the literature as capable of enzymatically hydrolysing glucosides,[113,114] a fact that can be of high interest in the case of other masked mycotoxins. Another important finding was that DON-3-Glc was not absorbed by the cellular monolayer in contrast to DON; 23% of DON was detected on the basolateral side, whereas less than 1% of DON-3-Glc crossed the Caco-2 monolayer. These findings for the bioavailability of DON-3-Glc are in line with an *in vivo* trial on rats, were DON-3-Glc was also described poorly bioavailable and was hydrolysed to DON during digestion.[115] Limited information is available regarding the intestinal absorption of DON, and the impact of the remaining mycotoxin in the intestinal lumen is still unknown. Another study evaluating absorption of FB_1 and its metabolites revealed that HFB_1, by losing its tricarballylic acid chain, is more bioavailable than its native toxin on a Caco-2 assay.[116] Finally, free ZEN in the lumen can be converted to its more oestrogenic phase I metabolites, αZEL and βZEL, by intestinal mucosal cells,[117] a fact that should trigger further investigations by applying masked ZEN into Caco-2 cells. Since these reports on DON and FB_1 are the only available information on *in vitro* bioavailabilty of masked mycotoxins they should not be regarded as being representative for all such compounds, for which further studies are needed.

6.5   Future Perspectives

The toxicological significance of masked mycotoxins is still obscure due to scarce toxicity data. The few *in vitro* studies concerning mycotoxin derivatives have mainly focused on their bioaccessibility, digestive fate and partially on ADME investigation, as previously discussed in this chapter. Therefore, rather little is known about their inherent toxicity and yet less about possible synergistic toxic effects with the parent compounds (*e.g.* co-exposure of DON and DON-3-Glc). Hence, it would appear relevant to introduce some basic *in vitro* toxicity studies in order to cast some light on these issues. A wide range of *in vitro* assays exist that could be applied to masked mycotoxins, either individually or as mixtures with their precursors. These assays have several different endpoints, which include effects on cell membrane integrity, cell energy (*e.g.* ATP production, mitochondrial effects), oxidative stress (*e.g.* GSH levels, red-ox status), the induction of apoptosis, cell proliferation (*e.g.* DNA replication) and protein synthesis. The determining factor in assay selection should primarily be the type of toxicological effects caused by the parent mycotoxins.

The common goal of every *in vitro* toxicity system has been to bridge the gap between the use of whole animals and the use of two-dimensional (2D) cellular monolayers that share little resemblance to living tissues. Target organ toxicity relying on 2D models is arguably far too simple and overlooks essential parameters, including interaction, communication and mechanical cues among different cell types within organs.[118] As a consequence, organ physiology is poorly represented. Moreover, traditionally used immortal cell lines have often lost characteristic features present in the primary cells of corresponding tissues, leading to abnormal responses to toxic compounds that cause serious deviations from *in vivo* studies. Immortal 2D cultured cell lines are gradually being replaced with primary cells in 2D cultures or co-cultures of different primary cells. The co-culture of various relevant primary cell types or use of tissue slices (*e.g.* precision-cut liver slices), although an improvement over single cell lines, suffers from rapid loss of characteristic features, limited lifespan and is problematic in high-throughput screening applications. One of the most promising solutions that has only recently been achieved, is the development of three-dimensional (3D) cell culture techniques that mimic native tissues much more realistically than ever before.[118] Even 3D high-throughput applications are nowadays possible, with up to 384-format arrays of 3D multicellular spheroids consisting of co-cultured primary cell types.[119] Figure 6.3 illustrates an example of liver microtissues that are formed in scaffold free 'hanging drop' culture plates, as co-cultures of rat hepatocytes and non-parenchymal cells, in comparison with actual human liver cells. These liver microtissues are viable for up to five weeks in culture, opening up new possibilities for *in vitro* long-term toxicity and biotransformation studies.

Recently, organ-on-chip systems have been developed that mimic organ-specific changes on microscale cell culture platforms. In these applications, several fluid chambers and channels are connected to each cell culture well, enabling controlled fluid flow and culture conditions. The use of tissue slices, cell lines, primary cells, co-cultures of different cell types and even stem cells is becoming possible in these models. The ultimate goal of such applications is to provide a human on-a-chip model, with cells from several organs cultured simultaneously on one microplatform where a fluidistic system would mimic the blood circulation *in vivo*.[120–122]

Besides cellular applications offering closer to *in vivo* experimental conditions, gene expression tests could also provide valuable information regarding toxicity, biotransformation and modes of action of masked mycotoxins. There are multiple ways to investigate gene expression, from microarrays producing large amounts of data (*e.g.* Affymetrix® GeneChips covering more than 30 000 genes) to studies of single genes or sets of genes from a single pathway, using real-time polymerase chain reaction methods (*e.g.* TaqMan® Gene Expression Assays) or Northern blotting. However, it should be noted that they are usually laborious and costly assays, and require specialised equipment. One important issue to keep in mind is that the outcome of gene expression studies based on *in vitro* assays

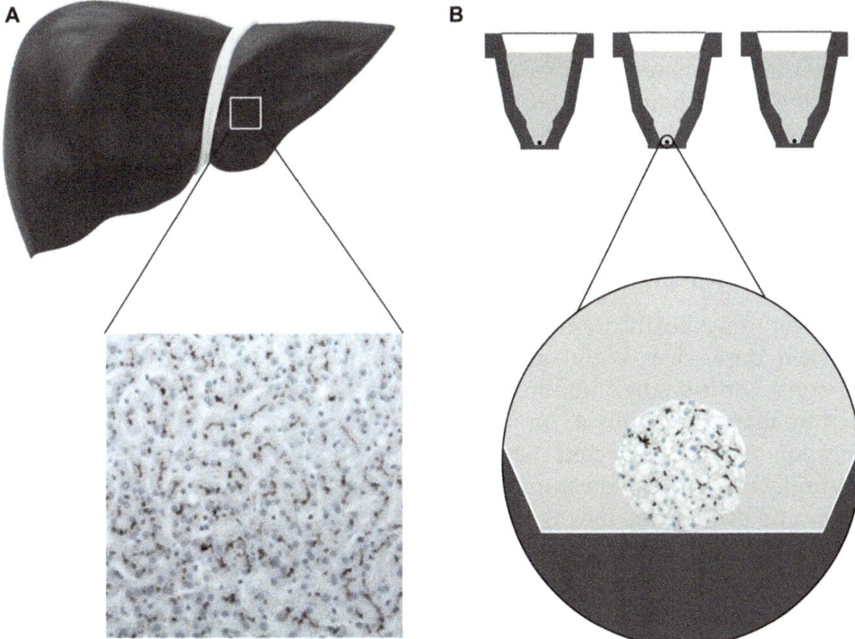

Figure 6.3 Immunohistochemistry of bile salt export pump (BSEP) in formalin-fixated paraffin embedded (FFPE) sections of intact, native human liver (A) and 3D human liver microtissues derived from primary hepatocytes in co-culture with non-parenchymal cells (NPCs) at culture day 21 (B). Images were taken with an Olympus BH2 microscope with an 10x objective. Images kindly provided by InSphero Inc., AG, Switzerland (© InSphero Inc., AG, Switzerland).

generally differs markedly from the *in vivo*. Boess *et al.*[123] conducted a comparative study on messenger ribonucleic acid (mRNA) expression profiles of *in vivo* studies with rodents *versus in vitro* setups, including precision-cut liver slices, primary hepatocytes in monolayer and sandwich culture, as well as hepatic cell lines. Precision-cut liver slices produced expression profiles closer to those obtained by animal studies, whereas immortal hepatic cell lines diverged the most.

The entrance of potentially harmful compounds into the cell is another key feature related to cytotoxicity. Very little has been published on the mechanisms of cell entry of mycotoxins or their masked forms. Gene expression data on the transporter protein expression of mycotoxin-treated cells and untreated cells can provide useful information on cell entry mechanisms. The functions of transport proteins can be also investigated with electrical measurements using inhibitors.[124] Unbound intracellular xenobiotic concentrations are pivotal in toxic interactions. Several methods exist to determine the intracellular levels of chemicals, such as mass spectrometry imaging,[125] but the unbound chemical concentrations within the cell, free to interact with cellular organelles and functions, are difficult to estimate. Mateus *et al.*[126] introduced a rather simple *in vitro* technique that combines measurements of intracellular binding (using equilibrium dialysis of cell homogenates) with steady-state measurements in living cells, to determine the unbound concentrations of several drugs within cells. All the above-mentioned techniques, together with the already utilised types of *in vitro* assays, could generate valuable knowledge on the emerging topic of masked mycotoxins.

6.6   Conclusion

For every mycotoxin excreted in the field or during storage, a number of masked derivatives could apparently emerge, co-existing with their precursors. Additionally, modified mycotoxins can be formed by fungal and mammalian metabolism, or interaction between mycotoxins and food components. It is expected that a vast quantity of these compounds exist in naturally contaminated commodities, with only a small fraction identified to date, and even fewer having been studied from a toxicological perspective. Nevertheless, the available data indicate low general toxicity, altered bioavailability and to some degree a possible risk of reactivation of masked mycotoxins during mammalian digestion. The release of precursor mycotoxins along the GIT may pose the main threat to human and animal health, depending on the digestive phase in which reactivation occurs. Early (enzymatic) de-conjugation might exert local toxicity to intestinal cells, while at the same time increasing the probability of native toxin absorption and transport into the systemic circulation. As digestion progresses, and ingested masked mycotoxins reach the large intestine chemically intact, the possibility of absorption decreases dramatically. Nonetheless, unpredictable consequences to the intestinal flora may unravel following microbial hydrolysis.

Based on current knowledge of the toxicity of masked mycotoxins *per se*, which in most cases seems to be lower than that of the parent compounds or metabolites thereof, most conjugation reactions could be considered as mitigation methods for avoiding the adverse effects of mycotoxins. The formation, toxicity and digestive stability of masked mycotoxins must, however, be further investigated from different angles before this mitigation concept can progress. Regardless of whether tests are performed to assess toxicity or their potential as detoxification agents, assays are needed that can rapidly and accurately provide scientists with such fundamental information in a cost-effective manner. Due to the numerous reasons presented in this chapter and the inclination towards alternative methods that reduce, refine and replace animal testing (3R principle), *in vitro* assays in general comprise the ideal way to examine toxicity of masked mycotoxins. *In vivo* studies should not be neglected though, particularly as a means to evaluate toxic significance on a systemic level. It can be concluded that there is presently too little toxicological information to enable the risk assessment of masked mycotoxins. Taking into account the enormous chemical diversity of masked mycotoxins and the lack of information concerning their toxicological relevance to human health, masked mycotoxins are a highly intriguing research topic from a toxicological viewpoint.

References

1. C. Dall'Asta, G. Galaverna, A. Dossena, S. Sforza and R. Marchelli, Masked mycotoxins and mycotoxin derivatives in food: the hidden menace, in *Mycotoxins in Food, Feed and Bioweapons*, ed. M. Rai and A. Varma, Springer, Berlin Heidelberg, 2010, vol. 22, pp. 385–397.
2. C. L. Broadhead and R. D. Combes, The current status of food additives toxicity testing and the potential for application of the three Rs, *ATLA, Altern. Lab. Anim.*, 2001, **29**, 471–485.
3. C. D. Klaassen, *Casarett and Doull's Toxicology: The Basic Science of Poisons*, The McGraw-Hill Companies Inc., New York, 7th edn, 2008.
4. K. K. Sinha and D. Bhatnagar, *Mycotoxins in Agriculture and Food Safety*, Marcel Dekker Inc., New York, 1998, pp. 183–254.
5. V. Betina, *Mycotoxins: Chemical, Biological, and Environmental Aspects*, Elsevier, London, 1989, pp. 114–150.
6. CAST, *Mycotoxins: Risks in Plant Animal, and Human Systems*, Task Force Report No. 139, Council for Agricultural Science and Technology, Iowa, 2003.
7. International Agency for Research on Cancer (IARC), Some naturally occurring substances: food items and constituents, heterocyclic aromatic amines and mycotoxins, in *IARC Monographs on the Evaluation of Carcinogenic Risks of Chemicals to Humans*, World Health Organization, Lyon, 1993, vol. 56, pp. 245–446.
8. R. Bhat, R. V. Rai and A. A. Karim, Mycotoxins in food and feed: present status and future concerns, *Compr. Rev. Food Sci. Food Saf.*, 2010, **9**, 57–81.
9. O. Rocha, K. Ansari and F. M. Doohan, Effects of trichothecene mycotoxins on eukaryotic cells: a review, *Food Addit. Contam.*, 2005, **22**, 369–378.
10. K. Peltonen, M. Jestoi and G. S. Eriksen, Health effects of moniliformin: a poorly understood *Fusarium* mycotoxin, *World Mycotoxin J.*, 2010, **3**, 403–414.
11. J. J. Pestka, Deoxynivalenol: mechanisms of action, human exposure, and toxicological relevance, *Arch. Toxicol.*, 2010, **84**, 663–679.
12. M. Maresca, From the gut to the brain: journey and pathophysiological effects of the food-associated trichothecene mycotoxin deoxynivalenol, *Toxins*, 2013, **5**, 784–820.
13. H. S. Hussein and J. M. Brasel, Toxicity, metabolism, and impact of mycotoxins on humans and animals, *Toxicology*, 2001, **167**, 101–134.
14. J. D. Miller and H. L. Trenholm, *Mycotoxins in Grain: Compounds Other Than Aflatoxin*, Eagan Press, Minnesota, 1994, pp. 487–539.

15. C. J. Kennedy and K. B. Tierney, Xenobiotic protection/resistance mechanisms in organisms, in *Environmental Toxicology*, ed. E. A. Laws, Springer, New York, 2013, vol. 23, pp. 689–722.

16. J. O. D. Coleman, M. M. A. Blake-Kalff and T. G. E. Davies, Detoxification of xenobiotics by plants: chemical modification and vacuolar compartmentation, *Trends Plant Sci.*, 1997, **2**, 144–151.

17. U. A. Boelsterli, *Mechanistic Toxicology: The Molecular Basis of How Chemicals Disrupt Biological Targets*, Taylor & Francis Group, Florida, 2nd edn, 2007, pp. 63–112.

18. F. Berthiller, R. Schuhmacher, G. Adam and R. Krska, Formation, determination and significance of masked and other conjugated mycotoxins, *Anal. Bioanal. Chem.*, 2009, **395**, 1243–1252.

19. D. M. Bartholomew, D. E. Van Dyk, S. M. C. Lau, D. P. O'Keefe, P. A. Rea and P. V. Viitanen, Alternate energy-dependent pathways for the vacuolar uptake of glucose and glutathione conjugates, *Plant Physiol.*, 2002, **130**, 1562–1572.

20. D. Bowles, E. K. Lim, B. Poppenberger and F. E. Vaistij, Glycosyltransferases of lipophilic small molecules, *Annu. Rev. Plant. Biol.*, 2006, **57**, 567–597.

21. S. S. Deshpande, *Handbook of Food Toxicology*, Marcel Dekker Inc., New York, 2002, pp. 75–118.

22. A. Zinedine, J. M. Soriano, J. C. Molto and J. Manes, Review on the toxicity, occurrence, metabolism, detoxification, regulations and intake of zearalenone: an oestrogenic mycotoxin, *Food Chem. Toxicol.*, 2007, **45**, 1–18.

23. A. Yiannikouris and J. P. Jouany, Mycotoxins in feeds and their fate in animals: a review, *Anim. Res.*, 2002, **51**, 81–99.

24. EU Directive 2003/15/EC of the European parliament and the council of 27 February 2003 amending council directive 76/768/EEC on the approximation of the laws of the member states relating to cosmetic products, *Off. J. Eur. Union*, 2003, **L66**, 26–35.

25. S. Coecke, O. Pelkonen, S. B. Leite, U. Bernauer, J. G. M. Bessems, F. Y. Bois, U. Gundert-Remy, G. Loizou, E. Testai and J. M. Zaldivar, Toxicokinetics as a key to the integrated toxicity risk assessment based primarily on non-animal approaches, *Toxicol. In Vitro*, 2013, **27**, 1570–1577.

26. S. Adler, D. Basketter, S. Creton, O. Pelkonen, J. van Benthem, V. Zuang, K. E. Andersen, A. Angers-Loustau, A. Aptula, A. Bal-Price, E. Benfenati, U. Bernauer, J. Bessems, F. Y. Bois, A. Boobis, E. Brandon, S. Bremer, T. Broschard, S. Casati, S. Coecke, R. Corvi, M. Cronin, G. Daston, W. Dekant, S. Felter, E. Grignard, U. Gundert-Remy, T. Heinonen, I. Kimber, J. Kleinjans, H. Komulainen, R. Kreiling, J. Kreysa, S. B. Leite, G. Loizou, G. Maxwell, P. Mazzatorta, S. Munn, S. Pfuhler, P. Phrakonkham, A. Piersma, A. Poth, P. Prieto, G. Repetto, V. Rogiers, G. Schoeters, M. Schwarz, R. Serafimova, H. Tahti, E. Testai, J. van Delft, H. van Loveren, M. Vinken, A. Worth and J. M. Zaldivar,

Alternative (non-animal) methods for cosmetics testing: current status and future prospects–2010, *Arch. Toxicol.*, 2011, **85**, 367–485.

27. F. A. Barile, *Principles of Toxicology Testing*, Taylor & Francis Group, Florida, 2008, pp. 147–290.

28. B. Kluger, C. Bueschl, M. Lemmens, F. Berthiller, G. Haubl, G. Jaunecker, G. Adam, R. Krska and R. Schuhmacher, Stable isotopic labelling-assisted untargeted metabolic profiling reveals novel conjugates of the mycotoxin deoxynivalenol in wheat, *Anal. Bioanal. Chem.*, 2013, **405**, 5031–5036.

29. M. Busman, S. M. Poling and C. M. Maragos, Observation of T-2 toxin and HT-2 toxin glucosides from *Fusarium sporotrichioides* by liquid chromatography coupled to tandem mass spectrometry (LC-MS/MS), *Toxins*, 2011, **3**, 1554–1568.

30. H. Nakagawa, K. Ohmichi, S. Sakamoto, Y. Sago, M. Kushiro, H. Nagashima, M. Yoshida and T. Nakajima, Detection of a new *Fusarium* masked mycotoxin in wheat grain by high-resolution LC–Orbitrap™ MS, *Food Addit. Contam., Part A*, 2011, **28**, 1447–1456.

31. G. Stacey, Current developments in cell culture technology, in *New Technologies for Toxicity Testing*, ed. M. Balls, R. D. Combes and N. Bhogal, Springer, New York, 2012, vol. 1, 1–13.

32. D. Acosta, E. M. B. Sorensen, D. C. Anuforo, D. B. Mitchell, K. Ramos, K. S. Santone and M. A. Smith, An *in vitro* approach to the study of target organ toxicity of drugs and chemicals, *In Vitro Cell. Dev. Biol.*, 1985, **21**, 495–504.

33. D. Zhou and Y. Qiu, *In vitro-in vivo* correlations of pharmaceutical dosage forms, in *Oral Bioavailability: Basic Principles, Advanced Concepts, and Applications*, ed. M. Hu and X. Li, John Wiley & Sons Inc., New Jersey, 2011, vol. 7, pp. 77–89.

34. J. V. Castell and M. J. Gomez-Lechon, *In Vitro Methods in Pharmaceutical Research*, Academic Press Inc., San Diego, 1997, pp. 1–54.

35. T. A. Kocarek, E. G. Schuetz and P. S. Guzelian, Expression of multiple forms of cytochrome P450 mRNAs in primary cultures of rat hepatocytes maintained on matrigel, *Mol. Pharmacol.*, 1993, **43**, 328–334.

36. P. Maurel, The use of adult human hepatocytes in primary culture and other *in vitro* systems to investigate drug metabolism in man, *Adv. Drug Delivery Rev.*, 1996, **22**, 105–132.

37. G. Eisenbrand, B. Pool-Zobel, V. Baker, M. Balls, B. J. Blaauboer, A. Boobis, A. Carere, S. Kevekordes, J. C. Lhuguenot, R. Pieters and J. Kleiner, Methods of *in vitro* toxicology, *Food Chem. Toxicol.*, 2002, **40**, 193–236.

38. K. Abado-Becognee, T. A. Mobio, R. Ennamany, F. Fleurat-Lessard, W. T. Shier, F. Badria and E. E. Creppy, Cytotoxicity of fumonisin B1: implication of lipid peroxidation and inhibition of protein and DNA syntheses, *Arch. Toxicol.*, 1998, **72**, 233–236.

39. E. E. Creppy, R. Roschenthaler and G. Dirheimer, Inhibition of protein synthesis in mice by ochratoxin A and its prevention by phenylalanine, *Food Chem. Toxicol.*, 1984, **22**, 883–886.

40. F. Berthiller, C. Crews, C. Dall'Asta, S. De Saeger, G. Haesaert, P. Karlovsky, I. P. Oswald, W. Seefelder, G. Speijers and J. Stroka, Masked mycotoxins: a review, *Mol. Nutr. Food Res.*, 2013, **57**, 165–186.

41. B. Poppenberger, F. Berthiller, D. Lucyshyn, T. Sieberer, R. Schuhmacher, R. Krska, K. Kuchler, J. Glossl, C. Luschnig and G. Adam, Detoxification of the *Fusarium* mycotoxin deoxynivalenol by a UDP-glucosyltransferase from *Arabidopsis thaliana*, *J. Biol. Chem.*, 2003, **278**, 47905–47914.

42. C. H. A. Snijders, Resistance in wheat to *Fusarium* infection and trichothecene formation, *Toxicol. Lett.*, 2004, **153**, 37–46.

43. A. Mesterhazy, Role of deoxynivalenol in aggressiveness of *Fusarium Graminearum* and *F. culmorum* and in resistance to *Fusarium* head blight, *Eur. J. Plant Pathol.*, 2002, **108**, 675–684.

44. M. Lemmens, U. Scholz, F. Berthiller, C. Dall'Asta, A. Koutnik, R. Schuhmacher, G. Adam, H. Buerstmayr, A. Mesterhazy, R. Krska and P. Ruckenbauer, The ability to detoxify the mycotoxin deoxynivalenol colocalizes with a major quantitative trait locus for *Fusarium* head blight resistance in wheat, *Mol. Plant–Microbe Interact.*, 2005, **18**, 1318–1324.

45. S. T. Tran and T. K. Smith, Conjugation of deoxynivalenol by *Alternaria alternata* (54028 NRRL), *Rhizopus microsporus var. rhizopodiformis* (54029 NRRL) and *Aspergillus oryzae* (5509 NRRL), *Mycotoxin Res.*, 2014, **30**, 47–53.

46. A. L. Boutigny, F. Richard-Forget and C. Barreau, Natural mechanisms for cereal resistance to the accumulation of *Fusarium* trichothecenes, *Eur. J. Plant Pathol.*, 2008, **121**, 411–423.

47. Joint FAO/WHO Expert Committee on Food Additives (JECFA), *Evaluation of Certain Contaminants in Food, WHO Technical Report Series No. 959*, World Health Organization, Geneva, 2011.

48. EFSA Panel on Contaminants in the Food Chain (CONTAM), Scientific opinion on the risks for public health related to the presence of zearalenone in food, *EFSA J.*, 2011, **9**(2197), 124.

49. J. W. Bennett and M. Klich, Mycotoxins, *Clin. Microbiol. Rev.*, 2003, **16**, 497–516.

50. H. A. C. Atkinson and K. Miller, Inhibitory effect of deoxynivalenol, 3-acetyldeoxynivalenol and zearalenone on induction of rat and human lymphocyte proliferation, *Toxicol. Lett.*, 1984, **23**, 215–221.

51. Z. Vlata, F. Porichis, G. Tzanakakis, A. Tsatsakis and E. Krambovitis, A study of zearalenone cytotoxicity on human peripheral blood mononuclear cells, *Toxicol. Lett.*, 2006, **165**, 274–281.

52. Joint FAO/WHO Expert Committee on Food Additives (JECFA), *Safety Evaluation of Certain Food Additives and Contaminants, WHO Food Additives Series No. 44*, World Health Organization, Geneva, 2000.

53. H. Malekinejad, R. F. Maas-Bakker and J. Fink-Gremmels, Bioactivation of zearalenone by porcine hepatic biotransformation, *Vet. Res.*, 2005, **36**, 799–810.

54. F. Berthiller, U. Werner, M. Sulyok, R. Krska, M. T. Hauser and R. Schuhmacher, Liquid chromatography coupled to tandem mass spectrometry (LC-MS/MS) determination of phase II metabolites of the mycotoxin zearalenone in the model plant *Arabidopsis thaliana*, *Food Addit. Contam.*, 2006, **23**, 1194–1200.
55. M. P. Kovalsky Paris, W. Schweiger, C. Hametner, R. Stuckler, G. J. Muehlbauer, E. Varga, R. Krska, F. Berthiller and G. Adam, Zearalenone-16-*O*-glucoside: a new masked mycotoxin, *J. Agric. Food Chem.*, 2014, **62**, 1181–1189.
56. H. Kamimura, Conversion of zearalenone to zearalenone glycoside by *Rhizopus sp.*, *Appl. Environ. Microbiol.*, 1986, **52**, 515–519.
57. S. el-Sharkawy and Y. Abul-Hajj, Microbial transformation of zearalenone, I. Formation of zearalenone-4-*O*-*β*-2-glucoside, *J. Nat. Prod.*, 1987, **50**, 520–521.
58. B. Poppenberger, F. Berthiller, H. Bachmann, D. Lucyshyn, C. Peterbauer, R. Mitterbauer, R. Schuhmacher, R. Krska, J. Glossl and G. Adam, Heterologous expression of *Arabidopsis* UDP-glucosyltransferases in *Saccharomyces cerevisiae* for production of zearalenone-4-*O*-glucoside, *Appl. Environ. Microbiol.*, 2006, **72**, 4404–4410.
59. J. Plasencia and C. J. Mirocha, Isolation and characterization of zearalenone sulfate produced by *Fusarium spp*, *Appl. Environ. Microbiol.*, 1991, **57**, 146–150.
60. P. M. Scott, Recent research on fumonisins: a review, *Food Addit. Contam., Part A*, 2012, **29**, 242–248.
61. C. Dall'Asta, M. Mangia, F. Berthiller, A. Molinelli, M. Sulyok, R. Schuhmacher, R. Krska, G. Galaverna, A. Dossena and R. Marchelli, Difficulties in fumonisin determination: the issue of hidden fumonisins, *Anal. Bioanal. Chem.*, 2009, **395**, 1335–1345.
62. C. Dall'Asta, G. Galaverna, G. Aureli, A. Dossena and R. Marchelli, A LC/MS/MS method for the simultaneous quantification of free and masked fumonisins in maize and maize-based products, *World Mycotoxin J.*, 2008, **1**, 237–246.
63. C. Dall'Asta, C. Falavigna, G. Galaverna, A. Dossena and R. Marchelli, *In vitro* digestion assay for determination of hidden fumonisins in maize, *J. Agric. Food Chem.*, 2010, **58**, 12042–12047.
64. W. T. Shier, The fumonisin paradox: a review of research on oral bioavailability of fumonisin B1, a mycotoxin produced by *Fusarium Moniliforme*, *J. Toxicol.*, 2000, **19**, 161–187.
65. M. Ruhland, G. Engelhardt and P. Wallnofer, Transformation of the mycotoxin ochratoxin A in artificially contaminated vegetables and cereals, *Mycotoxin Res.*, 1997, **13**, 54–60.
66. S. C. Gad, *Toxicology of the Gastrointestinal Tract*, Taylor & Francis Group, Florida, 2007.
67. C. H. M. Versantvoort, E. van de Kamp and C. J. M. Rompelberg, Development and applicability of an in vitro digestion model in assessing

the bioaccessibility of contaminants from food, *RIVM report 320102002*, 2004.

68. A. G. Oomen, C. J. M. Rompelberg, M. A. Bruil, C. J. G. Dobbe, D. P. K. H. Pereboom and A. J. A. M. Sips, Development of an *in vitro* digestion model for estimating the bioaccessibility of soil contaminants, *Arch. Environ. Contam. Toxicol.*, 2003, **44**, 281–287.

69. C. A. Gonzalez-Arias, S. Marin, V. Sanchis and A. J. Ramos, Mycotoxin bioaccessibility/absorption assessment using *in vitro* digestion models: a review, *World Mycotoxin J.*, 2013, **6**, 167–184.

70. E. F. A. Brandon, A. G. Oomen, C. J. M. Rompelberg, C. H. M. Versantvoort, J. G. M. van Engelen and A. J. A. M. Sips, Consumer product *in vitro* digestion model: bioaccessibility of contaminants and its application in risk assessment, *Regul. Toxicol. Pharmacol.*, 2006, **44**, 161–171.

71. A. Raiola, G. Meca, J. Manes and A. Ritieni, Bioaccessibility of deoxynivalenol and its natural co-occurrence with ochratoxin A and aflatoxin B1 in Italian commercial pasta, *Food Chem. Toxicol.*, 2012, **50**, 280–287.

72. B. Grenier and T. J. Applegate, Modulation of intestinal functions following mycotoxin ingestion: meta-analysis of published experiments in animals, *Toxins*, 2013, **5**, 396–430.

73. J. Parada and J. M. Aguilera, Food microstructure affects the bioavailability of several nutrients, *J. Food Sci.*, 2007, **72**, 21–32.

74. S. J. Hur, B. O. Lim, E. A. Decker and D. J. McClements, *In vitro* human digestion models for food applications, *Food Chem.*, 2011, **125**, 1–12.

75. C. H. M. Versantvoort, A. G. Oomen, E. van de Kamp, C. J. M. Rompelberg and A. J. A. M. Sips, Applicability of an *in vitro* digestion model in assessing the bioaccessibility of mycotoxins from food, *Food Chem. Toxicol.*, 2005, **43**, 31–40.

76. A. C. Guyton and J. E. Hall, Secretory functions of the alimentary tract. In: *Textbook of Medical Physiology*, Elsevier Inc., Philadelphia, 11th edn, 2006, vol. 64, pp. 791–807.

77. P. Rantonen, *Salivary Flow and Composition in Healthy and Diseased Adults*, University of Helsinki, Helsinki, 2003, pp. 12–32.

78. S. P. Humphrey and R. T. Williamson, A review of saliva: normal composition, flow, and function, *J. Prosthet. Dent.*, 2001, **85**, 162–169.

79. F. Kong and R. P. Singh, Modes of disintegration of solid foods in simulated gastric environment, *Food Biophys.*, 2009, **4**, 180–190.

80. J. Maldonado-Valderrama, P. Wilde, A. Macierzanka and A. Mackie, The role of bile salts in digestion, *Adv. Colloid Interface Sci.*, 2011, **165**, 36–46.

81. P. B. Eckburg, E. M. Bik, C. N. Bernstein, E. Purdom, L. Dethlefsen, M. Sargent, S. R. Gill, K. E. Nelson and D. A. Relman, Diversity of the human intestinal microbial flora, *Science*, 2005, **308**, 1635–1638.

82. P. Karlovsky, Biological detoxification of fungal toxins and its use in plant breeding, feed and food production, *Nat. Toxins*, 1999, 7, 1–23.

83. M. De Nijs, H. J. van den Top, L. Portier, G. Oegema, E. Kramer, H. P. van Egmond and L. A. P. Hoogenboom, Digestibility and absorption of deoxynivalenol-3-β-glucoside in *in vitro* models, *World Mycotoxin J.*, 2012, **5**, 319–324.

84. A. M. Aura, *In vitro* digestion models for dietary phenolic compounds, VTT Publications 575, VTT Technical Research Centre of Finland, Espoo, 2005, pp. 22–61.

85. N. Sewald, J. L. von Gleissenthall, M. Schuster, G. Muller and R. T. Aplin, Structure elucidation of a plant metabolite of 4-deoxynivalenol, *Tetrahedron: Asymmetry*, 1992, **3**, 953–960.

86. F. Berthiller, R. Krska, K. J. Domig, W. Kneifel, N. Juge, R. Schuhmacher and G. Adam, Hydrolytic fate of deoxynivalenol-3-glucoside during digestion, *Toxicol. Lett.*, 2011, **206**, 264–267.

87. A. Dall'Erta, M. Cirlini, M. Dall'Asta, D. Del Rio, G. Galaverna and C. Dall'Asta, Masked mycotoxins are efficiently hydrolyzed by human colonic microbiota releasing their aglycones, *Chem. Res. Toxicol.*, 2013, **26**, 305–312.

88. C. Falavigna, M. Cirlini, G. Galaverna and C. Dall'Asta, Masked fumonisins in processed food: co-occurrence of hidden and bound forms and their stability under digestive conditions, *World Mycotoxin J.*, 2012, **5**, 325–334.

89. E. De Angelis, L. Monaci, A. Mackie, L. Salt and A. Visconti, Bioaccessibility of T-2 and HT-2 toxins in mycotoxin contaminated bread models submitted to *in vitro* human digestion, *Innovative Food Sci. Emerging Technol.*, 2014, **22**, 248–256.

90. S. W. Gratz, G. Duncan and A. J. Richardson, The human fecal microbiota metabolizes deoxynivalenol and deoxynivalenol-3-glucoside and may be responsible for urinary deepoxy-deoxynivalenol, *Appl. Environ. Microbiol.*, 2013, **79**, 1821–1825.

91. E. L. Motta and P. M. Scott, Bioaccessibility of total bound fumonisin from corn flakes, *Mycotoxin Res.*, 2009, **25**, 229–232.

92. Q. Wu, V. Dohnal, K. Kuca and Z. Yuan, Trichothecenes: structure-toxic activity relationships, *Curr. Drug Metab.*, 2013, **14**, 641–660.

93. J. D. Hayes, D. J. Judah, L. I. McLellan and G. E. Neal, Contribution of the glutathione S-transferases to the mechanisms of resistance to aflatoxin B1, *Pharmacol. Ther.*, 1991, **50**, 443–472.

94. S. A. Gardiner, J. Boddu, F. Berthiller, C. Hametner, R. M. Stupar, G. Adam and G. J. Muehlbauer, Transcriptome analysis of the barley deoxynivalenol interaction: evidence for a role of glutathione in deoxynivalenol detoxification, *Mol. Plant–Microbe Interact.*, 2010, **23**, 962–976.

95. S. Danicke, K. Matthaus, P. Lebzien, H. Valenta, K. Stemme, K. H. Ueberschar, E. Razzazi-Fazeli, J. Bohm and G. Flachowsky, Effects of *Fusarium* toxin-contaminated wheat grain on nutrient turnover, microbial protein synthesis and metabolism of deoxynivalenol and zearalenone in the rumen of dairy cows, *J. Anim. Physiol. Anim. Nutr.*, 2005, **89**, 303–315.

96. M. J. Saint-Cyr, A. Perrin-Guyomard, P. Houee, J. G. Rolland and M. Laurentie, Evaluation of an oral subchronic exposure of deoxynivalenol on the composition of human gut microbiota in a model of human microbiota-associated rats, *PLoS One*, 2013, **8**, e80578.

97. I. Tenk, E. Fodor and Cs. Szathmary, The effect of pure *Fusarium* toxins (T-2, F-2, DAS) on the microflora of the gut and on plasma glucocorticoid levels in rat and swine, *Zentralbl. Bakteriol., Mikrobiol. Hyg., Ser. A*, 1982, **252**, 384–393.

98. K. Kulkarni and M. Hu, Caco-2 cell culture model for oral drug absorption, in *Oral Bioavailability: Basic Principles, Advanced Concepts, and Applications*, ed. M. Hu and X. Li, John Wiley & Sons Inc., New Jersey, 2011, vol. 27, pp. 431–442.

99. S. Yee, *In vitro* permeability across Caco-2 cells (colonic) can predict *in vivo* (small intestinal) absorption in man–fact or myth, *Pharm. Res.*, 1997, **14**, 763–766.

100. I. J. Hidalgo, T. J. Raub and R. T. Borchardt, Characterization of the human colon carcinoma cell line (Caco-2) as a model system for intestinal epithelial permeability, *Gastroenterology*, 1989, **96**, 736–749.

101. P. Artursson and J. Karlsson, Correlation between oral drug absorption in humans and apparent drug permeability coefficients in human intestinal epithelial (Caco-2) cells, *Biochem. Biophys. Res. Commun.*, 1991, **175**, 880–885.

102. P. Artursson, K. Palm and K. Luthman, Caco-2 monolayers in experimental and theoretical predictions of drug transport, *Adv. Drug Delivery Rev.*, 2001, **46**, 27–43.

103. S. Yamashita, T. Furubayashi, M. Kataoka, T. Sakane, H. Sezaki and H. Tokuda, Optimized conditions for prediction of intestinal drug permeability using Caco-2 cells, *Eur. J. Pharm. Sci.*, 2000, **10**, 195–204.

104. H. Lodish, A. Berk, P. Matsudaira, C. A. Kaiser, M. Krieger, M. P. Scott, L. Zipursky and J. Darnell, *Molecular Cell Biology*, W. H. Freeman and Company, New York, 5th edn, 2000, pp. 247–300.

105. D. W. Kufe, R. E. Pollock, R. R. Weichselbaum, R. C. Bast Jr, T. S. Gansler, J. F. Holland and E. Frei, *Holland-Frei Cancer Medicine*, BC Decker Inc., Ontario, 6th edn, 2003.

106. A. Leo, C. Hansch and D. Elkins, Partition coefficients and their uses, *Chem. Rev.*, 1971, **71**, 525–616.

107. T. J. Hou, W. Zhang, K. Xia, X. B. Qiao and X. J. Xu, ADME evaluation in drug discovery. 5. Correlation of Caco-2 permeation with simple molecular properties, *J. Chem. Inf. Comput. Sci.*, 2004, **44**, 1585–1600.

108. K. Gromadzka, A. Waskiewicz, J. Chelkowski and P. Golinski, Zearalenone and its metabolites: occurrence, detection, toxicity and guidelines, *World Mycotoxin J.*, 2008, **1**, 209–220.

109. D. Ringot, A. Chango, Y. J. Schneider and Y. Larondelle, Toxicokinetics and toxicodynamics of ochratoxin A, an update, *Chem. –Biol. Interact.*, 2006, **159**, 18–46.

110. A. Roth, K. Chakor, E. E. Creppy, A. Kane, R. Roschenthaler and G. Dirheimer, Evidence for an enterohepatic circulation of ochratoxin A in mice, *Toxicology*, 1988, **48**, 293–308.
111. K. A. Coddington, S. P. Swanson, A. S. Hassan and W. B. Buck, Enterohepatic circulation of T-2 toxin metabolites in the rat, *Drug Metab. Dispos.*, 1989, **17**, 600–605.
112. M. L. Biehl, D. B. Prelusky, G. D. Koritz, K. E. Hartin, W. B. Buck and H. L. Trenholm, Biliary excretion and enterohepatic cycling of zearalenone in immature pigs, *Toxicol. Appl. Pharmacol.*, 1993, **121**, 152–159.
113. A. Steensma, H. P. J. M. Noteborn, R. C. M. van der Jagt, T. H. G. Polman, M. J. B. Mengelers and H. A. Kuiper, Bioavailability of genistein, daidzein, and their glycosides in intestinal epithelial Caco-2 cells, *Environ. Toxicol. Pharmacol.*, 1999, **7**, 209–212.
114. J. Boyer, D. Brown and R. H. Liu, Uptake of quercetin and quercetin 3-glucoside from whole onion and apple peel extracts by Caco-2 cell monolayers, *J. Agric. Food Chem.*, 2004, **52**, 7172–7179.
115. V. Nagl, H. Schwartz, R. Krska, W. D. Moll, S. Knasmuller, M. Ritzmann, G. Adam and F. Berthiller, Metabolism of the masked mycotoxin deoxynivalenol-3-glucoside in rats, *Toxicol. Lett.*, 2012, **213**, 367–373.
116. F. Caloni, M. Spotti, G. Pompa, F. Zucco, A. Stammati and I. De Angelis, Evaluation of fumonisin B1 and its metabolites absorption and toxicity on intestinal cells line Caco-2, *Toxicon*, 2002, **40**, 1181–1188.
117. E. Pfeiffer, A. Kommer, J. S. Dempe, A. A. Hildebrand and M. Metzler, Absorption and metabolism of the mycotoxin zearalenone and the growth promotor zeranol in Caco-2 cells *in vitro*, *Mol. Nutr. Food Res.*, 2011, **55**, 560–567.
118. J. W. Haycock, 3D cell culture: a review of current approaches and techniques, in *3D Cell Culture Methods and Protocols*, ed. J. W. Haycock, Humana Press, New York, 2011, vol. 1, pp. 1–15.
119. Y. C. Tung, A. Y. Hsiao, S. G. Allen, Y. S. Torisawa, M. Ho and S. Takayama, High-throughput 3D spheroid culture and drug testing using a 384 hanging drop array, *Analyst*, 2011, **136**, 473–478.
120. C. Luni, E. Serena and N. Elvassore, Human-on-chip for therapy development and fundamental science, *Curr. Opin. Biotechnol.*, 2014, **25**, 45–50.
121. I. Meyvantsson and D. J. Beebe, Cell culture models in microfluidic systems, *Annu. Rev. Anal. Chem.*, 2008, **1**, 423–449.
122. C. W. Scott, M. F. Peters and Y. P. Dragan, Human induced pluripotent stem cells and their use in drug discovery for toxicity testing, *Toxicol. Lett.*, 2013, **219**, 49–58.
123. F. Boess, M. Kamber, S. Romer, R. Gasser, D. Muller, S. Albertini and L. Suter, Gene expression in two hepatic cell lines, cultured primary hepatocytes, and liver slices compared to the *in vivo* liver gene expression in rats: possible implications for toxicogenomics use of *in vitro* systems, *Toxicol. Sci.*, 2003, **73**, 386–402.

124. K. B. Goralski, G. Lou, M. T. Prowse, V. Gorboulev, C. Volk, H. Koepsell and D. S. Sitar, The cation transporters rOCT1 and rOCT2 interact with bicarbonate but play only a minor role for amantadine uptake into rat renal proximal tubules, *J. Pharmacol. Exp. Ther.*, 2002, **303**, 959–968.

125. X. Chu, K. Korzekwa, R. Elsby, K. Fenner, A. Galetin, Y. Lai, P. Matsson, A. Moss, S. Nagar, G. R. Rosania, J. P. F. Bai, J. W. Polli, Y. Sugiyama and K. L. R. Brouwer, and International Transporter Consortium, Intracellular drug concentrations and transporters: measurement, modeling, and implications for the liver, *Clin. Pharmacol. Ther.*, 2013, **94**, 126–141.

126. A. Mateus, P. Matsson and P. Artursson, Rapid measurement of intracellular unbound drug concentrations, *Mol. Pharmaceutics*, 2013, **10**, 2467–2478.

Animal Models for Masked Mycotoxin Studies

VERONIKA NAGL*[a,b] AND FRANZ BERTHILLER[a]

[a] Center for Analytical Chemistry, Department for Agrobiotechnolgy (IFA-Tulln), University of Natural Resources and Life Sciences, Vienna (BOKU), Konrad Lorenz Straße 20, 3430 Tulln, Austria;
[b] BIOMIN Research Center, BIOMIN Holding GmbH, Technopark 1, 3430 Tulln, Austria
*Email: veronika.nagl@biomin.net

7.1   Introduction

As part of their defense mechanisms, plants have the potential to metabolize mycotoxins, which results in the formation of masked mycotoxins. Enzymatic conjugation of mycotoxins to more polar compounds such as sugars, amino acids or sulfate groups has been verified for various mycotoxins, including deoxynivalenol (DON), zearalenone (ZEN), fumonisin B_1 (FB_1), ochratoxin A, nivalenol and T-2 and HT-2 toxins. So far, more than 30 different masked mycotoxins have been identified.[1,2] Due to sophisticated experimental methodologies and advances in analytical techniques, this number is constantly increasing.[3–5] However, information on the toxicological relevance of masked mycotoxins is limited and only available for a handful of these compounds. As a consequence, the health risks for humans and animals deriving from masked mycotoxins is currently largely unknown.

In principle, risk assessment of food/feed contaminants is based on two components: quantification of human/animal exposure to these substances (*via* food/feed and other routes) and characterization of the hazard (potential

Issues in Toxicology No. 24
Masked Mycotoxins in Food: Formation, Occurrence and Toxicological Relevance
Edited by Chiara Dall'Asta and Franz Berthiller

Published by the Royal Society of Chemistry, www.rsc.org

of these substances to cause adverse health effects).[6] While several surveys have addressed the occurrence of masked mycotoxins, reporting a frequent contamination of cereal grains, processed food and compound feeds,[7–11] the comparably low number of *in vivo* toxicity studies is related to the following factors. First, masked mycotoxins represent an emerging issue and toxicological experiments require considerable amounts of respective test substances. Both the synthesis and the purification of masked mycotoxins are demanding and time-consuming tasks. So far, only one masked myco-toxin is commercially available (deoxynivalenol-3-glucoside as a calibrant), underlining the limited accessibility of these compounds for toxicological evaluations.[2] Second, *in vivo* trials are cost intensive because of the technical equipment and personal expertise needed (*e.g.* suitable housing facilities or professional animal health care). Furthermore, animal studies have to be well designed to ensure an increase in knowledge compared to *in vitro* experiments. The latter represent a valuable tool for toxicity assessment and can even be superior to *in vivo* models (*e.g.* when it comes to repeatability of results or evaluation of specific dose–response relationships) (see Section 6.3). Moreover, they contribute greatly to the reduction, refinement and replacement of animal testing (the 3Rs principle), which is desirable due to both ethical and legal considerations. Yet only *in vivo* studies allow an investigation of complex systemic influences on toxicokinetics and toxicodynamics. Consequently, not only the interaction of different tissues or organ systems, but also recovery processes and chronic effects can be taken into account in these models.[12]

In general, toxicological testing comprises studies on biochemical effects (absorption, distribution, metabolism and elimination [ADME]), acute toxi-city, subacute/sub-chronic toxicity, chronic toxicity, reproductive functions, carcinogenicity or teratogenicity.[13] Masked mycotoxins pose a risk to human and animal health either by exerting biological activity on their own or by liberation of their respective parent toxin during digestion. Owing to this fact and to the characteristics of masked mycotoxin exposure (ingestion of low doses *via* feed/food[14]), studies on the subacute toxicity, chronic toxicity and ADME after oral administration of compounds are of special importance for hazard identification. Naturally, evaluations on the biological activity of masked mycotoxins focus on the mode of action of the respective parent toxin and address reproductive (ZEN) or anorectic effects (DON). For ADME studies on masked mycotoxins, biotransformation of applied substances over the course of phase I and phase II reactions (see Section 6.2.2) has to be taken into consideration. The broad range of potentially formed metabolites and their low concentrations in biological fluids, especially in blood, provide substantial challenges for analytics. Since urine and feces contain compar-ably high levels of (masked) mycotoxin metabolites and allow non-invasive sampling procedures, analysis of excreta is often used to estimate the bioavailability of (masked) mycotoxins.[15,16]

In the following, the available animal studies on the toxicological relevance of masked ZEN, masked DON and masked fumonisins will be

Table 7.1 *In vivo* studies on the toxicological relevance of masked mycotoxins.

Masked mycotoxin	Species	Toxin administration	Investigated parameters	Ref.
ZEN-14-Glc	Pig	Multiple oral doses, 600 µg animal^{-1} day^{-1} for 2 weeks	Metabolites in urine and feces	34
	Rat	Single oral dose, 25 µg animal^{-1}	Metabolites in gastrointestinal tract, serum and organs	36
ZEN-14-S	Rat	Single oral dose, 630 nmol animal^{-1}	Uterus enlargement	28
DON-3-Glc	Rat	Single oral dose, 3.1 mg kg^{-1} b.w.	Metabolites in urine and feces	54, 57
	Rat	Single oral dose, 25 µg animal^{-1}	Metabolites in gastrointestinal tract, serum and organs	36
	Pig	Single oral dose, 116 µg kg^{-1} b.w.; single intravenous dose, 15.5 µg kg^{-1} b.w.	Metabolites in urine and feces	65
	Mouse	Single oral dose, 2.5 mg kg^{-1} b.w.	Splenic cytokine and chemokine mRNA expression	67
	Mouse	Single oral doses, 2.5, 5, 10 mg kg^{-1} b.w.	Feed refusal	69
	Mouse	Single oral dose, 2.5 mg kg^{-1} b.w.	Feed refusal and levels of gut satiety hormones in plasma	69
	Mink	Single oral doses, 0.05, 0.25, 0.5, 1, 2 mg kg^{-1} b.w.	Emetic potency	69

presented (Table 7.1). Whenever possible, special emphasis will be put on the comparison of the effects of masked mycotoxins and their respective parent toxin. Other mycotoxin derivatives included in some of the mentioned studies (*e.g.* fungal toxins and synthetic analogues) will not be addressed in this context.

7.2   Animal Studies on Masked ZEN

ZEN, mainly produced by *Fusarium graminearum*, *F. crookwellense* and *F. culmorum*, is of low acute toxicity, but poses a risk to human and animal health due to its strong estrogenic effects.[17] ZEN acts as a full and partial agonist on estrogen receptors α and β, respectively, thereby causing functional and morphological alterations in reproductive organs. In pigs—the species regarded to be most sensitive to ZEN—clinical signs include ovarian atrophy, enlargement of the uterus, swelling of the vulva, decreased fertility and stillbirth.[18] ZEN is rapidly absorbed after oral administration.[19] In the course of phase I metabolism, ZEN is reduced to α- and β-zearalenol (αZEL and βZEL) by hydroxysteroid dehydrogenases, predominantly in intestinal or liver cells.[20,21] While αZEL has an even higher binding affinity to estrogen receptors than its parent toxin, βZEL on the other hand exerts lower estrogenic activity.[22] Thus, species-specific variations in the rate of α-hydroxylation may account for differences in the susceptibility towards ZEN among species.[23] Further reduction of αZEL and βZEL, resulting in the formation of α- and β-zearalanol (αZAL and βZAL), seems to have minor relevance *in vivo*, while glucuronidation of ZEN, αZEL and βZEL by uridine diphosphate (UDP)-glucuronyl transferases in liver and extrahepatic tissues (phase II metabolism) is quite prominent in most investigated species.[24–26] Thereafter, ZEN and its metabolites are excreted *via* bile (enterohepatic recycling) or urine.[18]

ZEN undergoes extensive biotransformation not only in mammals, but also in plants and fungi. For example, 17 different metabolites were detected after ZEN treatment of the model plant *Arabidopsis thaliana*, including zeralenone-14-*O*-β-D-glucopyranoside (ZEN-14-Glc), α-zearalenol-14-*O*-β-D-glucopyranoside (αZEL-14-Glc) and β-zearalenol-14-*O*-β-D-glucopyranoside (βZEL-14-Glc).[27] Recently, a second ZEN glucoside conjugate, zearalenone-16-*O*-β-D-glucopyranoside (ZEN-16-Glc), has been discovered after artificial infection of barley seeds.[4] Besides glucosylation, plants are capable of converting ZEN to zearalenone-14-sulfate (ZEN-14-S).[27] Yet this biotransformation pathway can also be observed in fungi,[28,29] and to some extent in mammals.[30,31] The occurrence of ZEN-14-Glc, αZEL-14-Glc, βZEL-14-Glc and ZEN-14-S has been demonstrated for compound feeds and various foodstuffs, such as bread, breakfast cereals or oatmeal, thus confirming an exposure of both farm animals and humans to masked forms of ZEN.[8,10,11]

To investigate the estrogenic activity of ZEN conjugates, ZEN-14-Glc and ZEN-14-S were tested for their ability to interact with human estrogen receptors. Both compounds, however, were found to be inactive in those binding assays.[1,32] Although the attachment of glucose or sulfate moieties seems to inhibit the direct estrogenic effects, possible health risks can still derive from the hydrolysis of masked forms of ZEN during mammalian digestion. An early *in vitro* study strengthened these concerns: after incubation of ZEN-14-S with the contents of different pig intestinal segments, release of ZEN was observed already in the duodenum.[33] Gareis *et al.* were the first to investigate the cleavage of ZEN-14-Glc under *in vivo* conditions.[34] A diet artificially contaminated with 395 $\mu g\,kg^{-1}$ ZEN-14-Glc was fed to a gilt for a duration of 14 days (corresponding to approximately 600 μg ZEN-14-Glc animal^{-1} day^{-1}). Excreta were collected daily and analyzed for ZEN-14-Glc, ZEN, αZEL and βZEL by high-performance liquid chromatography with fluorescence detection. While considerable levels of ZEN as well as αZEL were detected in urine and feces, the masked mycotoxin itself was not found. The authors therefore asserted complete hydrolysis of ZEN-14-Glc in the digestive tract of pigs. Yet liberation of ZEN was not associated with clinical signs of hyperestrogenism. Notably, comparable ZEN concentrations led to pronounced estrogenic effects in female pigs in a former study of the same research group.[35]

Similarly, ZEN-14-S showed reduced uterotropic activity when compared to its parent toxin in a rat uterus enlargement bioassay.[28] In this experiment, equimolar doses (630 nmol) of ZEN-14-S and ZEN were administered intragastrically to six Sprague–Dawley rats each. Although the average uterus weight of animals receiving ZEN-14-S was lower compared to those receiving ZEN, a significant estrogenic effect of the masked mycotoxin was nevertheless detectable.

More recently, an experiment investigated the metabolism of ZEN-14-Glc in more detail, thus providing valuable data for the elucidation of observed differences in the toxicity of ZEN and its masked forms *in vivo*. Veršilovskis *et al.* administered a single oral dose of 25 µg ^{13}C-ZEN and 25 µg ZEN-14-Glc to two male Wistar rats.[36] Blood samples (17 minutes after the treatment) as well as liver, kidney, bladder, spleen, lung, stomach, small intestine and large intestine samples (55 minutes after treatment) were collected. Concentrations of ZEN, ^{13}C-ZEN, αZEL, βZEL and the respective relevant glycosylated and glucuronidated forms were determined in the different matrices by a validated liquid chromatography–tandem mass spectrometry (LC-MS/MS) method. The majority of the administered ZEN-14-Glc, accounting for around 40% of the given dose, was found in the stomach and its contents. Importantly, considerable amounts of ZEN (approximately 18% of the administered dose) were found in the stomach, confirming *in vivo* hydrolysis of ZEN-14-Glc in the upper digestive tract. Since absorption of substances is generally assumed to be highest in proximal parts of the gastrointestinal tract, a significant uptake of ZEN after ZEN-14-Glc exposure could be deduced. However, neither in serum nor in the bladder (containing variable amounts of urine) were any of the investigated analytes detected. Moreover, minor amounts of ZEN-14-Glc were determined in the colon. Hence, the authors concluded that this masked mycotoxin is not completely hydrolyzed during digestion in rats. Although the relatively short time period between toxin administration and sample collection has to be taken into consideration, an incomplete liberation of ZEN could serve as a reasonable explanation for the reduced biological activity of ZEN-14-Glc in earlier studies.

For humans, cleavage of masked forms of ZEN has only been examined *in vitro*. As illustrated in Section 6.4.1.4, ZEN-14-Glc, ZEN-16-Glc and ZEN-14-S were rapidly hydrolyzed when incubated with fecal slurry samples. Microbial cleavage was accompanied by significant liberation of ZEN.[4,37] In contrast, ZEN-14-Glc and ZEN-14-S were stable under conditions mimicking the upper digestive tract.[37]

To sum up, the data clearly indicate that exposure to masked forms of ZEN can be of concern for human and animal health due to a release of their parent toxin during digestion. A relevant proportion of ZEN conjugates seems to be hydrolyzed already in the upper digestive tract, while microbial cleavage in distal parts of the intestine is most likely even more efficient.

Therefore, the European Food Safety Authority (EFSA) Panel on Contaminants in the Food Chain (CONTAM) used a practical approach and ascribed toxicity to the masked forms of ZEN as being similar to their parent toxin.[14] At the same time, the need for more information on the toxicity and bioavailability of masked ZEN was highlighted. For example, in sheep, a species with distinct ZEN susceptibility, rumen microbiota might facilitate complete hydrolysis of ZEN-14-Glc and ZEN-14-S already at an early stage of digestion. Hence, further studies relating the species-specific differences in the metabolism of masked forms of ZEN to the severity of reproductive disorders are warranted.

7.3 Animal Studies on Masked DON

DON, produced by various *Fusarium* spp., is one of the most frequently occurring mycotoxins worldwide.[38] Through binding to the 60S subunit of ribosomes, DON inhibits protein biosynthesis.[39] Furthermore, DON initiates a ribotoxic stress response. In the course of this process, activation of mitogen-activated protein kinases leads to a dose-dependent alteration in the transcription and mRNA stability of pro-inflammatory genes (cytokines, chemokines and other immune-related proteins) or, ultimately, apoptosis.[40] In addition to immunomodulation, DON affects gut integrity and causes gastro-intestinal symptoms like anorexia or emesis in experimental animals.[41,42] The underlying mechanisms for the anorectic effects are not completely characterized yet, but most likely involve modification of neuroendocrine and cytokine signaling.[43] Although the administration of high levels of DON causes shock-like death (oral LD_{50} in mice: $46 \, mg \, kg^{-1}$ body weight, b.w.),[44] the chronic effects after indigestion of comparable low doses, including growth retardation and enhanced susceptibility to infectious diseases,[41,45] are of far higher practical relevance. In humans, DON has been associated with episodes of gastro-enteritis and its capacity to alter the immune system is of major concern.[46]

Notable differences in the sensitivity to DON have been described between species. These are most likely related to species-specific variations in the *in vivo* metabolism of this mycotoxin.[41] In general, three major biotransformation pathways are known for DON in mammals (Figure 7.1). First, anaerobic ruminal or intestinal microbes facilitate the detoxification of DON to de-epoxy deoxynivalenol (DOM-1).[47–49] Second, absorbed DON and DOM-1 are conjugated to glucuronic acid as part of phase II metabolism, which results in the formation of deoxynivalenol-glucuronide (DON-GlcA) and de-epoxy deoxynivalenol-glucuronide (DOM-1-GlcA), respectively.[50,51] Different DON-GlcA isomers have been reported to occur *in vivo*, including deoxynivalenol-15-glucuronide (DON-15-GlcA), deoxynivalenol-3-glucuronide (DON-3-GlcA) and a third glucuronide with a so far unverified structure (deoxynivalenol-7-glucuronide or deoxynivalenol-8-glucuronide).[52–54] DON-15-GlcA is the predominant conjugate found in human urine and serves, in combination with urinary DON-3-GlcA and DON, as a validated biomarker for the assessment of human DON exposure.[52,55] The third metabolic pathway of DON, sulfonation, has long been neglected. Very recently, its

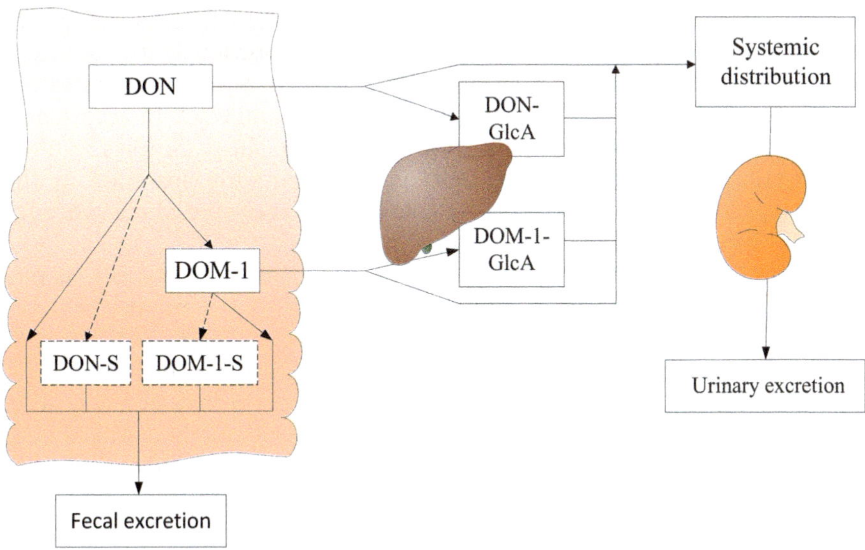

Figure 7.1   Main metabolic pathways of DON in monogastric animals.

importance for certain species has been highlighted.[56,57] Although the detailed mechanism has not been clarified yet, the formation of DON and DOM-1-sulfonate conjugates was suggested to occur in the gastrointestinal tract rather than after absorption of the parent compounds.[57]

Similar to mammals, plants are capable of metabolizing DON by various routes. Most prominently, glycosylation leads to the formation of deoxynivalenol-3-O-β-D-glucopyranoside (DON-3-Glc). A high prevalence of this masked mycotoxin has been confirmed for cereal grains, food products and animal compound feeds,[7,10] with levels of DON-3-Glc sometimes even exceeding those of the parent toxin.[58,59] In addition, biotransformation products related to glutathione metabolism (deoxynivalenol-glutathione, deoxynivalenol-*S*-cysteine, deoxynivalenol-*S*-cysteinyl-glycine) and sulfate conjugates were detected after artificial DON infection of wheat plants.[3,5] Yet the formation of these masked mycotoxins in naturally contaminated cereal crops has not been verified.

Due to its frequent occurrence and the existence of a commercially available analytical standard, a comparably higher number of *in vitro* and *in vivo* studies were performed on the toxicological relevance of DON-3-Glc. In the following sections, the potential health risks of DON-3-Glc, the liberation of DON during digestion and the possible direct toxicological effects will be addressed separately.

7.3.1   Stability of DON-3-Glc During Digestion

The journey of unravelling the fate of DON-3-Glc during digestion started with an *in vitro* study by Berthiller *et al.* in 2011.[60] Mimicking different stages of digestion, DON-3-Glc was found to be stable under acidic as well as enzymatic conditions. Yet several lactic acid bacteria, previously isolated from human gut, were capable of cleaving this masked mycotoxin, indicating hydrolysis of DON-3-Glc in distal parts of the intestinal tract. In accordance with this, incubation of DON-3-Glc with human fecal slurry samples resulted in liberation of DON.[37,61] Since DON is predominantly absorbed in the duodenum,[62] the location of DON-3-Glc hydrolysis is of special relevance for the increase of the total DON load of an individual.

Shortly thereafter, two reports on the metabolism of DON-3-Glc in rats became available. Nagl *et al.* administered equimolar amounts of DON (2.0 mg kg^{-1} b.w.) and DON-3-Glc (3.1 mg kg^{-1} b.w.) as single oral doses to six male Sprague–Dawley rats.[54] After each of the treatments, urine and feces were collected for 48 hours and analyzed for concentrations of DON-3-Glc, DON, DON-3-GlcA and DOM-1 by a validated LC-MS/MS method. Most of the applied DON-3-Glc was recovered in feces in the form of DON and DOM-1, indicating an extensive hydrolysis of this masked mycotoxin in the intestinal tract of rats. Only 3.7% of the given DON-3-Glc dose was found in urine, with DON, DON-3-GlcA and DOM-1 accounting for 1.3%, 1.2% and 0.7%, respectively. Although the presence of these analytes in urine provided evidence for the absorption and biotransformation of liberated DON *in vivo*, the overall absorption was markedly reduced compared to the DON treatment (by approximately a factor of 4). Notably, urinary DON-3-Glc represented only 0.3% of the administered dose. The low bioavailability of DON-3-Glc is in agreement with *in vitro* findings from De Nijs *et al.*, who demonstrated a lack of substantial absorption of this masked mycotoxin by human Caco-2 cells.[63]

Versilovskis *et al.* monitored the fate of DON-3-Glc in the intestinal tract of rats more closely.[36] In the experiment already described in Section 7.2, two male Wister rats additionally received 25 μg ^{13}C-DON and 25 μg DON-3-Glc by gavage. Sample analysis included LC-MS/MS determination of DON, ^{13}C-DON, DON-3-Glc, DOM-1 and the respective glucuronidated forms. After administration of DON-3-Glc, DON levels in the contents of the stomach accounted for only 2.3% of the given dose. Considering the results obtained for ZEN-14-Glc treatment, notable differences in the stability of masked mycotoxins in the upper digestive tract of rats seem to exist. Due to a sharp drop of DON-3-Glc concentrations in the small intestine, the authors proposed strong activity of enzymes like β-glucosidases. Yet the decline of DON-3-Glc was not accompanied by an increase of DON levels or metabolites thereof. With the exception of the urinary bladder (containing less than 0.1% of applied toxin dose), DON-3-Glc and DON were not detected in any other of the investigated organs, nor in serum. In total, approximately 50% of the administered DON-3-Glc dose could be recovered. This proportion is higher than that found in the experiment of Nagl *et al.*, in which 21% of the applied DON-3-Glc was found. Among others, the formation of unidentified DON-3-Glc metabolites was suggested as a possible reason for the lack of recovered toxin.

Driven by the discovery of novel DON metabolites,[56] a LC-MS/MS method was developed enabling the quantification of eight different DON-,

DOM-1- and DON-3-Glc-sulfonates in rat excreta.[57] Applying this analysis method to the samples collected in the experiment of Nagl *et al.*,[54] considerable amounts of sulfonate conjugates were found in the feces of DON-3-Glc-exposed animals. Specifically, fecal DON-, DOM-1- and DON-3-Glc-sulfonates together accounted for 47% of the administered toxin dose. Although DON-sulfonate 2 was found to be the major metabolite, fecal DON-3-Glc-sulfonate 2 represented 9.7% of the given dose. Hence, a limited proportion of orally administered DON-3-Glc was not cleaved, but rather directly biotransformed. Since only marginal amounts of sulfonate-conjugates were detected in urine, a low absorption of these metabolites can be assumed. These findings clearly highlight the continuous advances in analytical techniques and their importance for ADME studies on (masked) mycotoxins.

Metabolism of DON underlies considerable species-specific variations.[41] In this regard, humans and pigs share major similarities (*e.g.* high bioavailability of DON or negligible formation of DOM-1).[55,64] As a consequence, findings on the DON-3-Glc metabolism in pigs are presumed to have special relevance in terms of extrapolation of data to humans. In a follow-up experiment by Nagl *et al.*, equimolar amounts of DON (75 μg kg^{-1} b.w.) and DON-3-Glc (116 μg kg^{-1} b.w.) were administered orally to four male piglets on days 3 and 9, respectively.[65] In addition, the masked mycotoxin was administered intravenously on day 13 (15.5 μg kg^{-1} b.w.). After each of the treatments, urine and feces was collected for 24 hours in two indicated time periods (0–8 and 8–24 hours post-dosing) and analyzed for concentrations of DON, DON-3-GlcA, DON-15-GlcA, DON-3-Glc and DOM-1 by LC-MS/MS. The majority of orally administered DON-3-Glc was excreted via urine (40.3%), with DON-3-Glc, DON, DON-3-GlcA, DON-15-GlcA and DOM-1 accounting for 2.6%, 21.6%, 3.4%, 6.8% and 5.9% of the given dose, respectively. In contrast, only negligible toxin amounts (1.8%) were found in feces. While the primary excretion route of this masked mycotoxin seems to differ between species, the marginal levels of urinary DON-3-Glc and the presence of DON (and metabolites thereof) in urine are in line with results obtained for rats. Thus, the low bioavailability of unchanged DON-3-Glc, as well as its cleavage following oral exposure, are most likely common features of mammals. Compared to the DON treatment (84.8%), the urinary excretion of the masked mycotoxin and its metabolites was reduced by a factor of approximately 2 (Figure 7.2). Preliminary analyses revealed no significant formation of sulfonate metabolites in pigs (H. Schwartz-Zimmermann, personal communication). However, comparison of DON and DON-3-Glc treatment indicated a delayed excretion of the masked mycotoxin. Thus, the authors pointed out that the increase of the total DON burden deriving from hydrolysis of DON-3-Glc may be underestimated by 24 hour sampling in pigs. Following intravenous DON-3-Glc administration, the vast majority of the masked mycotoxin was excreted unchanged via urine, indicating a lack of substantial DON-3-Glc hydrolysis after absorption from the gastrointestinal tract.

Combining the information gained from both *in vitro* studies and animal experiments, DON-3-Glc itself seems to be only poorly absorbed.

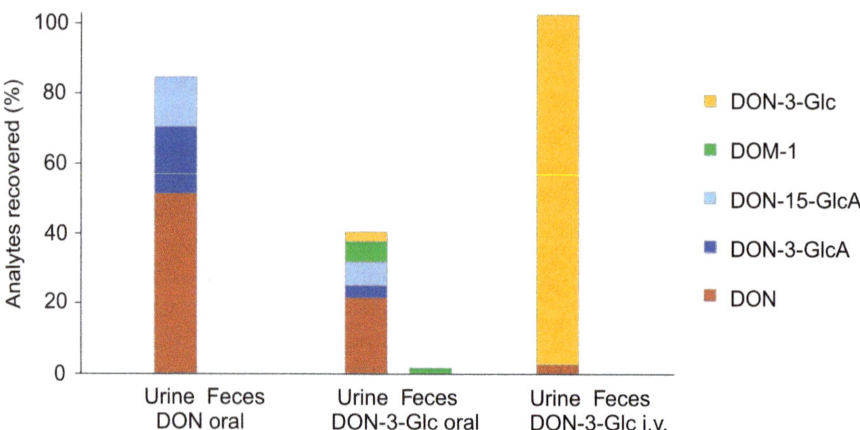

Figure 7.2   Excretion of DON, DON-3-Glc and their metabolites in the urine and feces of piglets after single oral (0.25 μmol kg^{-1} b.w.) or intravenous (0.03 μmol kg^{-1} b.w.) toxin administration (adapted from Nagl *et al.*[65]).

Furthermore, the cleavage of DON-3-Glc during digestion is most likely carried out in distal parts of the intestine, thus resulting in a decreased and/ or delayed absorption of liberated DON. However, nothing is known so far concerning the bioavailability of DON-3-Glc after chronic exposure (which may increase, as observed for DON[16]), especially via naturally contaminated grains. For risk assessment of DON-3-Glc, studies clarifying this aspect will be crucial.

7.3.2   Biological Activity of DON-3-Glc

Based on the *in vitro* findings by Poppenberger *et al.*, who demonstrated a markedly reduced activity of DON-3-Glc on protein translation in wheat ribosomes,[66] a low toxicity for this masked mycotoxin has been assumed. Although further *in vitro* studies reported a reduced cytotoxicity of DON-3-Glc (see Section 6.3.2), *in vivo* studies on the toxicological effects of DON-3-Glc were lacking for a long time.

   Recently, two studies by Wu *et al.* were published, evaluating the potential of DON-3-Glc to induce activation of the innate immune system and to evoke anorexia and emesis, respectively. In the first study, groups of six female B6C3F1 mice were orally gavaged with 2.5 mg kg^{-1} b.w. of DON and DON-3-Glc, respectively.[67] After 2 and 6 hours, splenic mRNA expression of cytokines (IL-1β, IL-6 and TNF-α) and chemokines (CXCL-2, CCL-2 and CCL-7) was measured by quantitative real-time polymerase chain reaction. While DON dramatically increased all investigated parameters (predominantly at the first sampling time point), only moderate elevation of IL-1β and IL-6 mRNA expression was observed in the DON-3-Glc treatment group. Moreover, mRNA expression of TNF-α, CXCL-2, CCL-2 and CCL-7 was completely unaffected by the masked mycotoxin. These results are of special relevance because splenic cytokine mRNA expression was shown to predict cytokine protein levels in the periphery,[67,68] and the latter are discussed as one of the trigger factors for DON-induced anorexia.[40] However, the capability of DON-3-Glc to induce anorexia and, subsequently, emesis was specifically investigated in a follow-up study describing a set of three consecutive animal experiments. First, groups of five female B6C3F1 mice were orally dosed with 0, 2.5, 5 and 10 mg kg^{-1} b.w. of DON or DON-3-Glc.[69] Thereafter, pre-weighted food pellets were provided to the animals and food intake was measured at six time points between 0.5 and 16 hours post-exposure. Compared to the DON treatment, which caused the rapid onset of anorexia with a >80% decrease in cumulative food intake, the feed reduction after DON-3-Glc administration was reduced and delayed. Cumulative feed intake from 1 to 6 hours after treatment was decreased by 42–70% for all DON-3-Glc dose levels.

In the second experiment, the authors examined the role of gut satiety hormones in the onset of DON-3-Glc-induced anorexia in female B6C3F1 mice. To this end, two groups of animals ($n = 6$) were orally gavaged with 2.5 mg kg^{-1} b.w. DON and DON-3-Glc, respectively. In addition to observation of the feed consumption, plasma concentrations of cholecystokinin (CCK) and peptide YY$_{3-36}$ (PYY$_{3-36}$) were determined by a competitive enzyme immunoassay at 0, 0.5, 2 and 6 hours after dosing. In line to the findings of the previous experiment, DON-3-Glc caused a delayed feed refusal in animals, which was most distinctive between 2 and 6 hours after toxin exposure. In addition, feed refusal could indeed be related to increased levels of gut satiety hormones for both DON and DON-3-Glc. Compared to the control, DON-3-Glc induced significant elevation of plasma CCK and PYY$_{3-36}$ levels at 2 and 6 hours after treatment, respectively.

In the third experiment, a mink model was used to assess the emetic potency of DON-3-Glc. Two female minks each were orally dosed with 0, 0.05, 0.25, 0.5, 1 and 2 mg kg^{-1} b.w. DON-3-Glc and subsequently monitored for retching and emesis over a time period of 3 hours. While DON-3-Glc doses up to 1 mg kg^{-1} b.w. had no effect, administration of 2 mg kg^{-1} b.w. evoked an emetic response in one of the animals. In contrast to the time patterns of feed refusal, onset of emesis was rapid (within 30 minutes). Despite the limited number of animals and dosage levels, the data facilitated the first-time estimation of toxicological parameters for DON-3-Glc, including the no observed adverse effect level (NOAEL; 1 mg kg^{-1}) and the lowest observed adverse effect level (LOAEL; 2 mg kg^{-1}) for oral emetic effects. Compared to previous results obtained for DON,[70] the NOAEL and LOAEL for DON-3-Glc were reduced by factors of 100 and 40, respectively.

To sum up, DON-3-Glc has been demonstrated to induce anorexia as well as emesis in mammals, albeit to a lower extent than its parent toxin. Gut satiety hormones rather than pro-inflammatory cytokines seem to mediate DON-3-Glc-induced anorexia. The delayed biological response of DON-3-Glc, particularly observed for onset of feed refusal, was assumed to stem from DON-3-Glc transformation to DON and/or its intestinal transit.[69]

Although these suggestions fit nicely with results gained from metabolism studies, experiments directly correlating the toxic effects of DON-3-Glc to the release of DON are lacking so far. In addition, DON-3-Glc may have a negative impact on gut integrity and permeability (either directly or indirectly *via* liberation of DON). Hence, further studies are needed to elucidate the biological activity of this masked mycotoxin in sufficient detail.

7.4   Animal Studies on Masked Fumonisins

Fumonisins are a group of mycotoxins predominantly produced by various *Fusarium* spp. or *Aspergillus niger*.[71,72] Of numerous fumonisin analogues that have been identified so far, FB_1 is the most relevant one in terms of prevalence and toxicity.[73] The range of adverse health effects induced by FB_1 is broad, including liver and kidney toxicity, genotoxicity, neurotoxicity, teratogenicity and carcinogenicity.[74] The sensitivity and primary target organ of FB_1 vary between species, strain and sex. In farm animals, exposure to FB_1 can cause specific diseases, namely equine leukoencephalomalacia and porcine pulmonary edema.[75] In humans, FB_1 has been associated with esophageal cancer and listed as a class 2B carcinogen by the International Agency for Research and Cancer.[76,77] In addition, FB_1 has been implicated as a possible risk factor for neural tube defects.[78]

Fumonisin toxicity is mainly based on disruption of the sphingolipid metabolism. Due to its structural similarity with free sphingoid bases, FB_1 acts as competitive inhibitor of the enzyme ceramide synthase, which catalyzes the formation of ceramide from fatty acids and sphinganine (Sa) or sphingosine (So).[79] As a consequence, FB_1 leads to intracellular accumulation of Sa and to a lesser extent So, which is reflected by an elevated Sa/So ratio. In addition, increased levels of sphinganine-1-phosphate, sphingosine-1-phosphate, 1-deoxysphinganine and 1-deoxysphingosine become evident.[73] These alterations, determined in plasma, urine or various tissues, serve as specific biomarkers for FB_1 exposure (biomarkers of effect) in animal models.[80] In contrast, urinary FB_1 is currently regarded as the most suitable biomarker for assessment of human fumonisin exposure.[55] However, determination of FB_1 in urine (and blood) requires highly sensitive analytical methods, as FB_1 is very poorly absorbed.[75] Also, fumonisins do not undergo substantial metabolism *in vivo*. So far, only marginal formations of partially ($pHFB_1a$ and $pHFB_1b$) and fully hydrolyzed FB_1 (HFB_1), most likely realized by gut microbiota, have been described.[81,82]

Despite its poor bioavailability, toxic effects were observed even after ingestion of low levels of FB_1, a phenomenon designated as the 'fumonisin paradox'.[83] As one possible explanation, the presence of unknown fumonisin derivatives in feed and food was hypothesized. Indeed, in the last decade, it became obvious that a significant proportion of fumonisins escapes routine detection due to modification of FB_1 *in planta* (masked fumonisins) or during food processing.[84,85] In the former case, FB_1 is either covalently linked to plant components, such as starch or fatty acids, or physically entrapped within macromolecules, *e.g.* proteins (non-covalent binding).[14,84] Although the chemical structures of these so-called matrix-associated fumonisins have not yet been elucidated, indirect analysis methods revealed the predominant formation of non-covalently bound masked fumonisins.[86] Since changes in the pH value or activity of certain enzymes can lead to the declustering of supramolecular structures, it is reasonable to assume that physically entrapped FB_1 is released during mammalian digestion, thus increasing the total toxin burden of an individual.[86,87] As recently highlighted by CONTAM, this scenario is of special concern for toddlers and other children.[14] Today, knowledge on the bioaccessibility of matrix-associated fumonisins remains is based on *in vitro* experiments, whereas dedicated *in vivo* studies addressing the metabolism and toxicity of these masked mycotoxins are completely absent so far. Mostly, this fact can be attributed the unknown structure and, subsequently, the unavailability of purified compounds. As a consequence, these obstacles have to be overcome before the conduction of sound *in vivo* studies on matrix-associated fumonisins is feasible.

The situation is quite different for covalently bound fumonisins deriving from food processing. In several *in vivo* studies, the metabolism and toxicity of thermal reaction products of FB_1 with reducing sugars N-(1-deoxy-D-fructos-1-yl) fumonisin B_1 (NDF-FB) or N-(carboxymethyl) fumonisin B_1 (NCM-FB) has been evaluated. It should be stressed that these toxins are referred to as modified mycotoxins rather than masked mycotoxins,[88] and are therefore only addressed briefly here. Although a previous study reported stability of NDF-FB under conditions imitating the upper digestive tract,[87] recent *in vitro* findings suggest partial cleavage of this modified mycotoxin during digestion.[89] In accordance with this, liberation of FB_1 from orally administered NDF-FB was proposed for rats.[82] Compared to its parent toxin, FB_1-glucose/fructose adducts and NCM-FB were shown to exhibit lower toxicity in rats, mice and swine.[82,90–92] The marked signs of toxicity after application of FB_1-glucose/fructose adducts that were observed in some studies (*e.g.* development of porcine pulmonary edema[92]) may be related to certain levels of unreacted FB_1 in applied toxin solutions. Distinct impurities impair the clear differentiation as to whether observed effects are caused by modified toxins, the release of FB_1 from modified toxins or unreacted FB_1. Hence, due to animal welfare considerations as well as the validity of the gained results, suitable methods for compound characterization, purification and subsequent analysis should be developed.

7.5   Conclusion

In vivo studies on the toxicological relevance of masked mycotoxins are limited, not only concerning the number of overall performed experiments, but also in terms of different application routes, toxin doses or used species. Nevertheless, the first valuable insights into the metabolism and biological activity of these compounds have been gained. Masked forms of ZEN and

DON were shown to evoke toxic effects in animals (*e.g.* uterus enlargement, feed refusal or emesis). Compared to the respective parent toxin, masked mycotoxins seem to possess lower toxicity *in vivo*. It has not been clarified in sufficient detail yet whether the observed adverse health effects derive from the direct biological activity of masked mycotoxins, from the release of the parent toxin during digestion or from a combination thereof. So far, *in vitro* toxicity assays and *in vivo* ADME studies suggest that hydrolysis of masked mycotoxins represents the major health risk. Yet the localization of cleavage as well as the rate of hydrolysis seems to vary between individual compounds and different species, thereby affecting the proportion of totally absorbed toxin. Hence, future studies need to investigate factors influencing the ADME of masked mycotoxins (*e.g.* the individual or species-specific composition of the gut microflora). In addition, attention has to be paid to the structure elucidation of as-yet unidentified metabolites of (masked) mycotoxins formed *in vivo*.

Another important aspect that has not been addressed concerns the effects arising from chronic exposure to masked mycotoxins, especially *via* naturally (co-)contaminated feed. Although the data point to a low bioavailability of masked mycotoxins, prolonged exposure may increase the absorption rates (*e.g.* by alteration of the gut integrity). However, such animal studies require considerable amounts of test substances. Therefore, challenges regarding compound synthesis and purification have to be met. The latter is also essential to assess the toxicity of further as-yet unevaluated or uncharacterized masked mycotoxins.

References

1. F. Berthiller, R. Schuhmacher, G. Adam and R. Krska, Formation, determination and significance of masked and other conjugated mycotoxins, *Anal. Bioanal. Chem.*, 2009, **395**, 1243–1252.
2. F. Berthiller, C. Crews, C. Dall'Asta, S. D. Saeger, G. Haesaert, P. Karlovsky, I. P. Oswald, W. Seefelder, G. Speijers and J. Stroka, Masked mycotoxins: A review, *Mol. Nutr. Food Res.*, 2013, **57**, 165–186.
3. B. Kluger, C. Bueschl, M. Lemmens, F. Berthiller, G. Häubl, G. Jaunecker, G. Adam, R. Krska and R. Schuhmacher, Stable isotopic labelling-assisted untargeted metabolic profiling reveals novel conjugates of the mycotoxin deoxynivalenol in wheat, *Anal. Bioanal. Chem.*, 2013, **405**, 5031–5036.
4. M. P. Kovalsky Paris, W. Schweiger, C. Hametner, R. Stuckler, G. J. Muehlbauer, E. Varga, R. Krska, F. Berthiller and G. Adam, Zearalenone-16-O-glucoside: a new masked mycotoxin, *J. Agric. Food Chem.*, 2014, **62**, 1181–1189.
5. B. Warth, P. Fruhmann, G. Wiesenberger, B. Kluger, B. Sarkanj, M. Lemmens, C. Hametner, J. Fröhlich, G. Adam, R. Krska and R. Schuhmacher, Deoxynivalenol-sulfates: identification and

quantification of novel conjugated (masked) mycotoxins in wheat, *Anal. Bioanal. Chem.*, 2015, **407**, 1033–1039.

6. European Food Safety Authority (EFSA), Risk assessment of contaminants in food and feed, *EFSA J.*, 2012, **10**, 1004.

7. F. Berthiller, R. Corradini, C. Dall'Asta, R. Marchelli, M. Sulyok, R. Krska, G. Adam and R. Schuhmacher, Occurrence of deoxynivalenol and its 3-β-D-glucoside in wheat and maize, *Food Addit. Contamin., A*, 2009, **26**, 507–511.

8. O. Vendl, C. Crews, S. MacDonald, R. Krska and F. Berthiller, Occurrence of free and conjugated Fusarium mycotoxins in cereal-based food, *Food Addit. Contam., Part A*, 2010, **27**, 1148–1152.

9. A. Malachova, Z. Dzuman, Z. Veprikova, M. Vaclavikova, M. Zachariasova and J. Hajslova, Deoxynivalenol, deoxynivalenol-3-glucoside, and enniatins: the major mycotoxins found in cereal-based products on the Czech market, *J. Agric. Food Chem.*, 2011, **59**, 12990–12997.

10. M. de Boevre, J. Diana di Mavungu, S. Landschoot, K. Audenaert, M. Eeckhout, P. Maene, G. Haesaert and S. de Saeger, *World Mycotoxin J.*, 2012, **5**, 207–219.

11. M. de Boevre, L. Jacxsens, C. Lachat, M. Eeckhout, J. D. di Mavungu, K. Audenaert, P. Maene, G. Haesaert, P. Kolsteren, B. de Meulenaer and S. de Saeger, Human exposure to mycotoxins and their masked forms through cereal-based foods in Belgium, *Toxicol. Lett.*, 2013, **218**, 281–292.

12. H. Spielmann In-Vitro-Methoden, in *Lehrbuch der Toxikologie* ed. H. Marquardt and S. Schäfer, BI-Wissenschaftlicher Verlag, Leipzig, 1994.

13. F. C. Lu, *Basic Toxicology: Fundamentals, Target Organs, and Risk Assessment*, Taylor & Francis, London, 1996.

14. EFSA Panel on Contaminants in the Food Chain, Scientific Opinion on the risks for human and animal health related to the presence of modified forms of certain mycotoxins in food and feed, *EFSA J.*, 2014, **12**, 3916.

15. P. Galtier, Biological fate of mycotoxins in animals, *Rev. Med. Vet.*, 1998, **149**, 549–554.

16. T. Goyarts and S. Dänicke, Bioavailability of the Fusarium toxin deoxynivalenol (DON) from naturally contaminated wheat for the pig, *Toxicol. Lett.*, 2006, **163**, 171–182.

17. A. Zinedine, J. M. Soriano, J. C. Molto and J. Manes, Review on the toxicity, occurrence, metabolism, detoxification, regulations and intake of zearalenone: an oestrogenic mycotoxin, *Food Chem. Toxicol.*, 2007, **45**, 1–18.

18. J. Fink-Gremmels and H. Malekinejad, Clinical effects and biochemical mechanisms associated with exposure to the mycoestrogen zearalenone, *Anim. Feed Sci. Technol.*, 2007, **137**, 326–341.

19. T. Kuiper-Goodman, P. Scott and H. Watanabe, Risk assessment of the mycotoxin zearalenone, *Regul. Toxicol. Pharmacol.*, 1987, **7**, 253–306.

20. M. Olsen and K. Kiessling, Species differences in zearalenone-reducing activity in subcellular fractions of liver from female domestic animals, *Acta Pharmacol. Toxicol.*, 1983, **52**, 287–291.
21. M. Olsen, H. Pettersson, K. Sandholm, A. Visconti and K. Kiessling, Metabolism of zearalenone by sow intestinal mucosa *in vitro*, *Food Chem. Toxicol.*, 1987, **25**, 681–683.
22. D. Fitzpatrick, C. Picken, L. Murphy and M. Buhr, Measurement of the relative binding affinity of zearalenone, alpha-zearalenol and beta-zearalenol for uterine and oviduct estrogen receptors in swine, rats and chickens: an indicator of estrogenic potencies, *Comp. Biochem. Physiol., C: Comp. Pharmacol.*, 1989, **94**, 691–694.
23. H. Malekinejad, R. Maas-Bakker and J. Fink-Gremmels, Species differences in the hepatic biotransformation of zearalenone, *Vet. J.*, 2006, **172**, 96–102.
24. E. Pfeiffer, A. Hildebrand, H. Mikula and M. Metzler, Glucuronidation of zearalenone, zeranol and four metabolites *in vitro*: formation of glucuronides by various microsomes and human UDP-glucuronosyltransferase isoforms, *Mol. Nutr. Food Res.*, 2010, **54**, 1468–1476.
25. M. Kleinova, P. Zöllner, H. Kahlbacher, W. Hochsteiner and W. Lindner, Metabolic profiles of the mycotoxin zearalenone and of the growth promoter zeranol in urine, liver, and muscle of heifers, *J. Agric. Food Chem.*, 2002, **50**, 4769–4776.
26. P. Zöllner, J. Jodlbauer, M. Kleinova, H. Kahlbacher, T. Kuhn, W. Hochsteiner and W. Lindner, Concentration levels of zearalenone and its metabolites in urine, muscle tissue, and liver samples of pigs fed with mycotoxin-contaminated oats, *J. Agric. Food Chem.*, 2002, **50**, 2494–2501.
27. F. Berthiller, U. Werner, M. Sulyok, R. Krska, M. T. Hauser and R. Schuhmacher, Liquid chromatography coupled to tandem mass spectrometry (LC-MS/MS) determination of phase II metabolites of the mycotoxin zearalenone in the model plant *Arabidopsis thaliana*, *Food Addit. Contam.*, 2006, **23**, 1194–1200.
28. J. Plasencia and C. J. Mirocha, Isolation and characterization of zearalenone sulfate produced by *Fusarium spp*, *Appl. Environ. Microbiol.*, 1991, **57**, 146–150.
29. A. Brodehl, A. Möller, H. Kunte, M. Koch and R. Maul, Biotransformation of the mycotoxin zearalenone by fungi of the genera Rhizopus and Aspergillus, *FEMS Microbiol. Lett.*, 2014, **359**, 124–130.
30. C. Mirocha, S. Pathre and T. Robison, Comparative metabolism of zearalenone and transmission into bovine milk, *Food Cosmet. Toxicol.*, 1981, **19**, 25–30.
31. M. Metzler, E. Pfeiffer and A. Hildebrand, Zearalenone and its metabolites as endocrine disrupting chemicals, *World Mycotoxin J.*, 2010, **3**, 385–401.
32. B. Poppenberger, F. Berthiller, H. Bachmann, D. Lucyshyn, C. Peterbauer, R. Mitterbauer, R. Schuhmacher, R. Krska, J. Glössl and G. Adam, Heterologous expression of Arabidopsis UDP-glucosyltransferases in

Saccharomyces cerevisiae for production of zearalenone-4-O-glucoside, *Appl. Environ. Microbiol.*, 2006, **72**, 4404–4410.

33. M. Gareis, Maskierte Mykotoxine, *Übers. Tierernähr*, 1994, **22**, 104–113.
34. M. Gareis, J. Bauer, J. Thiem, G. Plank, S. Grabley and B. Gedek, Cleavage of zearalenone-glycoside, a "masked" mycotoxin, during digestion in swine, *J. Vet. Med., Ser. B*, 1990, **37**, 236–240.
35. J. Bauer, K. Heinritzi, M. Gareis and B. Gedek, Veränderungen am Genitaltrakt des weiblichen Schweines nach Verfütterung praxisrelevanter Zearalenonmengen, *Tierärztl. Prax*, 1987, **15**, 33–36.
36. A. Veršilovskis, J. Geys, B. Huybrechts, E. Goossens, S. De Saeger and A. Callebaut, Simultaneous determination of masked forms of deoxynivalenol and zearalenone after oral dosing in rats by LC-MS/MS, *World Mycotoxin J.*, 2012, **5**, 303–318.
37. A. Dall'Erta, M. Cirlini, M. Dall'Asta, D. Del Rio, G. Galaverna and C. Dall'Asta, Masked mycotoxins are efficiently hydrolyzed by human colonic microbiota releasing their aglycones, *Chem. Res. Toxicol.*, 2013, **26**, 305–312.
38. I. Rodrigues and K. Naehrer, Prevalence of mycotoxins in feedstuffs and feed surveyed worldwide in 2009 and 2010, *Phytopathol. Mediterr.*, 2012, **51**, 175–192.
39. K. C. Ehrlich and K. W. Daigle, Protein synthesis inhibition by 8-oxo-12,13-epoxytrichothecenes, *Biochim. Biophys. Acta*, 1987, **923**, 206–213.
40. J. J. Pestka, Deoxynivalenol: mechanisms of action, human exposure, and toxicological relevance, *Arch. Toxicol.*, 2010, **84**, 663–679.
41. J. J. Pestka, Deoxynivalenol: Toxicity, mechanisms and animal health risks, *Anim. Feed Sci. Technol.*, 2007, **137**, 283–298.
42. A. P. Bracarense, J. Lucioli, B. Grenier, G. Drociunas Pacheco, W. D. Moll, G. Schatzmayr and I. P. Oswald, Chronic ingestion of deoxynivalenol and fumonisin, alone or in interaction, induces morphological and immunological changes in the intestine of piglets, *Br. J. Nutr.*, 2012, **107**, 1776–1786.
43. B. M. Flannery, W. Wu and J. J. Pestka, Characterization of deoxynivalenol-induced anorexia using mouse bioassay, *Food Chem. Toxicol.*, 2011, **49**, 1863–1869.
44. T. Yoshizawa and N. Morooka, Studies on the toxic substances in infected cereals; acute toxicities of new trichothecene mycotoxins: deoxynivalenol and its monoacetate, *J. Food Hyg. Soc. Jpn.*, 1974, **15**, 261–269.
45. G. Antonissen, A. Martel, F. Pasmans, R. Ducatelle, E. Verbrugghe, V. Vandenbroucke, S. Li, F. Haesebrouck, F. Van Immerseel and S. Croubels, The impact of Fusarium mycotoxins on human and animal host susceptibility to infectious diseases, *Toxins*, 2014, **6**, 430–452.
46. J. J. Pestka and A. T. Smolinski, Deoxynivalenol: Toxicology and potential effects on humans, *J. Toxicol. Environ. Health, Part B*, 2005, **8**, 39–69.
47. R. R. King, R. E. McQueen, D. Levesque and R. Greenhalgh, Transformation of deoxynivalenol (vomitoxin) by rumen microorganisms, *J. Agric. Food Chem.*, 1984, **32**, 1181–1183.

48. S. Swanson, C. Helaszek, W. Buck, H. Rood and W. Haschek, The role of intestinal microflora in the metabolism of trichothecene mycotoxins, *Food Chem. Toxicol.*, 1988, **26**, 823–829.

49. G. Sundstøl Eriksen, H. Pettersson and T. Lundh, Comparative cytotoxicity of deoxynivalenol, nivalenol, their acetylated derivatives and de-epoxy metabolites, *Food Chem. Toxicol.*, 2004, **42**, 619–624.

50. X. Wu, P. Murphy, J. Cunnick and S. Hendrich, Synthesis and characterization of deoxynivalenol glucuronide: its comparative immunotoxicity with deoxynivalenol, *Food Chem. Toxicol.*, 2007, **45**, 1846–1855.

51. V. M. T. Lattanzio, M. Solfrizzo, A. De Girolamo, S. N. Chulze, A. M. Torres and A. Visconti, LC-MS/MS characterization of the urinary excretion profile of the mycotoxin deoxynivalenol in human and rat, *J. Chromatogr. B*, 2011, **879**, 707–715.

52. B. Warth, M. Sulyok, F. Berthiller, R. Schuhmacher and R. Krska, New insights into the human metabolism of the Fusarium mycotoxins deoxynivalenol and zearalenone, *Toxicol. Lett.*, 2013, **220**, 88–94.

53. S. Uhlig, L. Ivanova and C. K. Faeste, Enzyme-assisted synthesis and structural characterization of the 3-, 8-, and 15-glucuronides of deoxynivalenol, *J. Agric. Food Chem.*, 2013, **61**, 2006–2012.

54. V. Nagl, H. Schwartz, R. Krska, W. D. Moll, S. Knasmüller, M. Ritzmann, G. Adam and F. Berthiller, Metabolism of the masked mycotoxin deoxynivalenol-3-glucoside in rats, *Toxicol. Lett.*, 2012, **213**, 367–373.

55. P. C. Turner, B. Flannery, C. Isitt, M. Ali and J. Pestka, The role of biomarkers in evaluating human health concerns from fungal contaminants in food, *Nutr. Res. Rev.*, 2012, **25**, 162–179.

56. D. Wan, L. Huang, Y. Pan, Q. Wu, D. Chen, Y. Tao, X. Wang, Z. Liu, J. Li and L. Wang, Metabolism, distribution, and excretion of deoxynivalenol with combined techniques of radiotracing, high-performance liquid chromatography ion trap time-of-flight mass spectrometry, and online radiometric detection, *J. Agric. Food Chem.*, 2013, **62**, 288–296.

57. H. E. Schwartz-Zimmermann, C. Hametner, V. Nagl, V. Slavik, W. Moll and F. Berthiller, Deoxynivalenol (DON) sulfonates as major DON metabolites in rats: from identification to biomarker method development, validation and application, *Anal. Bioanal. Chem.*, 2014, **406**, 7911–7924.

58. J. J. Sasanya, C. Hall and C. Wolf-Hall, Analysis of deoxynivalenol, masked deoxynivalenol, and Fusarium graminearum pigment in wheat samples, using liquid chromatography-UV-mass spectrometry, *J. Food Prot.*, 2008, **71**, 1205–1213.

59. E. Varga, A. Malachova, H. Schwartz, R. Krska and F. Berthiller, Survey of deoxynivalenol and its conjugates deoxynivalenol-3-glucoside and 3-acetyl-deoxynivalenol in 374 beer samples, *Food Addit. Contam., Part A*, 2013, **30**, 137–146.

60. F. Berthiller, R. Krska, K. J. Domig, W. Kneifel, N. Juge, R. Schuhmacher and G. Adam, Hydrolytic fate of deoxynivalenol-3-glucoside during digestion, *Toxicol. Lett.*, 2011, **206**, 264–267.

61. S. W. Gratz, G. Duncan and A. J. Richardson, The human fecal microbiota metabolizes deoxynivalenol and deoxynivalenol-3-glucoside and may be responsible for urinary deepoxy-deoxynivalenol, *Appl. Environ. Microbiol.*, 2013, **79**, 1821–1825.

62. S. Dänicke, H. Valenta and S. Döll, On the toxicokinetics and the metabolism of deoxynivalenol (DON) in the pig, *Arch. Anim. Nutr.*, 2004, **58**, 169–180.

63. M. De Nijs, H. Van den Top, L. Portier, G. Oegema, E. Kramer, H. Van Egmond and L. Hoogenboom, Digestibility and absorption of deoxynivalenol-3-ß-glucoside in *in vitro* models, *World Mycotoxin J*, 2012, **5**, 319–324.

64. S. Dänicke and U. Brezina, Kinetics and metabolism of the Fusarium toxin deoxynivalenol in farm animals: consequences for diagnosis of exposure and intoxication and carry over, *Food Chem. Toxicol.*, 2013, **60**, 58–75.

65. V. Nagl, B. Woechtl, H. E. Schwartz-Zimmermann, I. Hennig-Pauka, W. Moll, G. Adam and F. Berthiller, Metabolism of the masked mycotoxin deoxynivalenol-3-glucoside in pigs, *Toxicol. Lett.*, 2014, **229**, 190–197.

66. B. Poppenberger, F. Berthiller, D. Lucyshyn, T. Sieberer, R. Schuhmacher, R. Krska, K. Kuchler, J. Glössl, C. Luschnig and G. Adam, Detoxification of the Fusarium mycotoxin deoxynivalenol by a UDP-glucosyltransferase from *Arabidopsis thaliana*, *J. Biol. Chem.*, 2003, **278**, 47905–47914.

67. W. Wu, K. He, H. Zhou, F. Berthiller, G. Adam, Y. Sugita-Konishi, M. Watanabe, A. Krantis, T. Durst and H. Zhang, Effects of oral exposure to naturally-occurring and synthetic deoxynivalenol congeners on proinflammatory cytokine and chemokine mRNA expression in the mouse, *Toxicol. Appl. Pharmacol.*, 2014, **278**, 107–115.

68. Islam and J. J. Pestka, LPS priming potentiates and prolongs proinflammatory cytokine response to the trichothecene deoxynivalenol in the mouse, *Toxicol. Appl. Pharmacol.*, 2006, **211**, 53–63.

69. W. Wu, H. R. Zhou, S. J. Bursian, X. Pan, J. E. Link, F. Berthiller, G. Adam, A. Krantis, T. Durst and J. J. Pestka, Comparison of anorectic and emetic potencies of deoxynivalenol (vomitoxin) to the plant metabolite deoxynivalenol-3-glucoside and synthetic deoxynivalenol derivatives EN139528 and EN139544, *Toxicol. Sci.*, 2014, **142**, 167–181.

70. W. Wu, M. A. Bates, S. J. Bursian, J. E. Link, B. M. Flannery, Y. Sugita-Konishi, M. Watanabe, H. Zhang and J. J. Pestka, Comparison of emetic potencies of the 8-ketotrichothecenes deoxynivalenol, 15-acetyldeoxynivalenol, 3-acetyldeoxynivalenol, fusarenon X, and nivalenol, *Toxicol. Sci.*, 2013, **131**, 279–291.

71. J. P. Rheeder, W. F. Marasas and H. F. Vismer, Production of fumonisin analogs by *Fusarium* species, *Appl. Environ. Microbiol.*, 2002, **68**, 2101–2105.

72. J. C. Frisvad, J. Smedsgaard, R. A. Samson, T. O. Larsen and U. Thrane, Fumonisin B_2 production by *Aspergillus niger*, *J. Agric. Food Chem.*, 2007, **55**, 9727–9732.

73. K. Voss and R. Riley, Fumonisin toxicity and mechanism of action: overview and current perspectives, *Food Saf.*, 2013, **1**, 49–69.
74. H. Stockmann-Juvala and K. Savolainen, A review of the toxic effects and mechanisms of action of fumonisin B_1, *Hum. Exp. Toxicol.*, 2008, **27**, 799–809.
75. K. Voss, G. Smith and W. Haschek, Fumonisins: Toxicokinetics, mechanism of action and toxicity, *Anim. Feed Sci. Technol.*, 2007, **137**, 299–325.
76. G. Sun, S. Wang, X. Hu, J. Su, T. Huang, J. Yu, L. Tang, W. Gao and J. Wang, Fumonisin B_1 contamination of home-grown corn in high-risk areas for esophageal and liver cancer in China, *Food Addit. Contam.*, 2007, **24**, 181–185.
77. International Agency on Research on Cancer (IARC), *IARC Monogr. Eval. Carcinog. Risks Hum.*, 2002, **82**, 301–306.
78. L. Suarez, M. Felkner, J. D. Brender, M. Canfield, H. Zhu and K. A. Hendricks, Neural tube defects on the Texas-Mexico border: what we've learned in the 20 years since the Brownsville cluster, *Birth Defects Res., Part A*, 2012, **94**, 882–892.
79. A. H. Merrill Jr, G. van Echten, E. Wang and K. Sandhoff, Fumonisin B_1 inhibits sphingosine (sphinganine) *N*-acyltransferase and *de novo* sphingolipid biosynthesis in cultured neurons *in situ*, *J. Biol. Chem.*, 1993, **268**, 27299–27306.
80. R. T. Riley and K. A. Voss, in *Determining Mycotoxins and Mycotoxigenic Fungi in Food and Feed*, ed. S. De Saeger, Woodhead Publishing LTD, Cambridge, 2011.
81. G. S. Shephard, P. G. Thiel, E. W. Sydenham and M. E. Savard, Fate of a single dose of ^{14}C-labelled fumonisin B_1 in vervet monkeys, *Nat. Toxins*, 1995, **3**, 145–150.
82. I. Hahn, V. Nagl, H. E. Schwartz-Zimmermann, E. Varga, C. Schwarz, V. Slavik, N. Reisinger, A. Malachova, M. Cirlini, S. Generotti, C. Dall'Asta, R. Krska, W. Moll and F. Berthiller, Effects of orally administered fumonisin B_1 (FB_1), partially hydrolysed FB_1, hydrolysed FB_1 and *N*-(1-deoxy-D-fructos-1-yl) FB_1 on the sphingolipid metabolism in rats, *Food Chem. Toxicol.*, 2014, **76**, 11–18.
83. W. T. Shier, The fumonisin paradox: a review of research on oral bioavailability of fumonisin B_1, a mycotoxin produced by *Fusarium moniliforme*, *Toxin Rev.*, 2000, **19**, 161–187.
84. C. Dall'Asta, M. Mangia, F. Berthiller, A. Molinelli, M. Sulyok, R. Schuhmacher, R. Krska, G. Galaverna, A. Dossena and R. Marchelli, Difficulties in fumonisin determination: the issue of hidden fumonisins, *Anal. Bioanal. Chem.*, 2009, **395**, 1335–1345.
85. H. Humpf and K. A. Voss, Effects of thermal food processing on the chemical structure and toxicity of fumonisin mycotoxins, *Mol. Nutr. Food Res.*, 2004, **48**, 255–269.
86. C. Dall'Asta, C. Falavigna, G. Galaverna, A. Dossena and R. Marchelli, *In vitro* digestion assay for determination of hidden fumonisins in maize, *J. Agric. Food Chem.*, 2010, **58**, 12042–12047.

87. C. Falavigna, M. Cirlini, G. Galaverna and C. Dall'Asta, Masked fumonisins in processed food: co-occurrence of hidden and bound forms and their stability under digestive conditions, *World Mycotoxin J.*, 2012, **5**, 325–334.

88. M. Rychlik, H. Humpf, D. Marko, S. Dänicke, A. Mally, F. Berthiller, H. Klaffke and N. Lorenz, Proposal of a comprehensive definition of modified and other forms of mycotoxins including "masked" mycotoxins, *Mycotoxin Res.*, 2014, **30**, 197–205.

89. M. Cirlini, I. Hahn, E. Varga, M. Dall'Asta, C. Falavigna, L. Calani, F. Berthiller, D. Del Rio and C. Dall'Asta, Hydrolysed fumonisin B_1 and N-(deoxy-D-fructos-1-yl)-fumonisin B_1: stability and catabolic fate under simulated human gastrointestinal conditions, *Int. J. Food Sci. Nutr.*, 2015, **66**, 98–103.

90. P. C. Howard, L. H. Couch, R. E. Patton, R. M. Eppley, D. R. Doerge, M. I. Churchwell, M. M. Marques and C. V. Okerberg, Comparison of the toxicity of several fumonisin derivatives in a 28-day feeding study with female $B6C3F_1$ mice, *Toxicol. Appl. Pharmacol.*, 2002, **185**, 153–165.

91. H. Liu, Y. Lu, J. S. Haynes, J. E. Cunnick, P. Murphy and S. Hendrich, Reaction of fumonisin with glucose prevents promotion of hepatocarcinogenesis in female F344/N rats while maintaining normal hepatic sphinganine/sphingosine ratios, *J. Agric. Food Chem.*, 2001, **49**, 4113–4121.

92. G. Fernández-Surumay, G. D. Osweiler, M. J. Yaeger, C. C. Hauck, S. Hendrich and P. A. Murphy, Glucose reaction with fumonisin B_1 partially reduces its toxicity in swine, *J. Agric. Food Chem.*, 2004, **52**, 7732–7739.

Detoxification Strategies for Mycotoxins in Plant Breeding

PETR KARLOVSKY

Georg-August-University Goettingen, Molecular Phytopathology and Mycotoxin Research, Grisebachstrasse 6, D-37077 Goettingen, Germany
Email: pkarlov@gwdg.de

8.1 Fungal Toxins in Plant Diseases

Food production is constrained by pathogens and pests that attack growing plants, and harvested crops are vulnerable to spoilage and storage pests. Fungi are one of the causes of yield losses and quality deterioration in plant production. Fungal inoculum enters the food chain while crops are growing in the field or in a greenhouse and during storage, transport and processing of harvested plant commodities. Living plants are colonized by parasitic fungi, which cause plant diseases and are therefore pathogens, and by endophytes, which do not cause visible disease symptoms and may in certain situations even increase the fitness of crop plants. Endophytes may impair crop quality if they produce undesirable metabolites that accumulate in harvested plant organs. Harvested commodities consist of living tissue that has a limited capacity to counteract colonization by spoilage fungi and of dead material that is even less protected from spoilage.

Many fungal pathogens and all spoilage fungi produce mycotoxins. While the growth of fungi in stored products can be efficiently prevented by technical means such as drying and cooling, it is impossible to completely protect living plants from infection with pathogenic fungi in the field. Mycotoxin exposure *via* food produced in developed countries is therefore

Issues in Toxicology No. 24
Masked Mycotoxins in Food: Formation, Occurrence and Toxicological Relevance
Edited by Chiara Dall'Asta and Franz Berthiller
© The Royal Society of Chemistry 2016
Published by the Royal Society of Chemistry, www.rsc.org

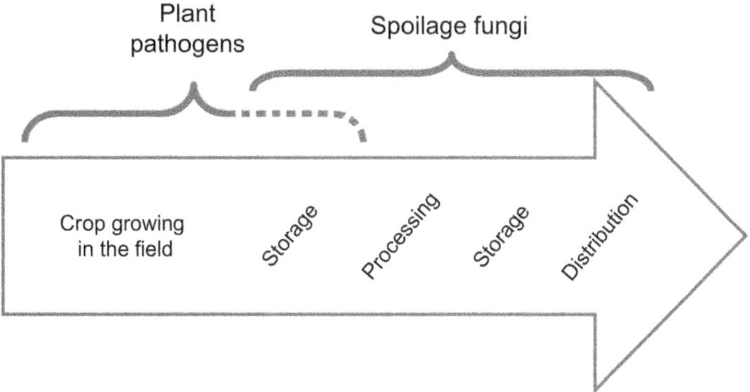

Figure 8.1 Origin of mycotoxins in food and feedstuff.

primarily caused by toxins produced by pathogens and accumulating in living plants in the field. Mycotoxin accumulation during storage jeopardizes food safety in countries with low standards of agricultural practices. In countries with high standards of food safety, major sources of alimentary mycotoxin exposure are imported food products (Figure 8.1).

This chapter will focus on toxins accumulating in living crop plants colonized by fungal pathogens. Toxic fungal metabolites have been a focus of plant breeders for two reasons. Firstly, ingestion of mycotoxins threatens the health of consumers and farm animals. Mycotoxin contamination diminishes the commercial value of crop products: when the mycotoxin content exceeds legal limits for food and feeds, the crop can only be used in biogas and ethanol production or destroyed, and blending contaminated batches with clean material is prohibited by law in Europe.

The second reason why breeders are interested in fungal toxins is that these compounds often act as virulence or pathogenicity factors. Virulence factors facilitate colonization of host plants and aggravate disease symptoms; pathogens that have lost the ability to produce a virulence factor are still capable of causing disease. Pathogenicity factors are necessary for a disease to occur; loss of the ability to produce a pathogenicity factor for a particular host renders the fungus apathogenic on that host. Virulence and pathogenicity factors interfere with plant defense responses, facilitate further colonization of plant tissue, aggravate disease symptoms and enhance the negative impacts of fungal infection on yield and quality. When susceptible plants are treated with purified fungal toxins acting as pathogenicity factors, these metabolites often cause symptoms similar to those triggered by their producers. Such metabolites are often called pathotoxins. By definition, virulence and pathogenicity factors do not have to be phytotoxic, but only a few virulence factors and no pathogenicity factors that are not phytotoxic are known. This fact, however, may reflect the easiness of identification of phytotoxic metabolites in fractionated fungal extracts rather than real-world situations. On the other hand, metabolites of non-pathogenic fungi may be toxic to plants, and plant pathogens may produce metabolites that exert phytotoxicity *in vitro* but are not involved in disease etiology. Every fungal metabolite that is soluble in water is likely to be phytotoxic under specific conditions.

Certain fungal metabolites facilitating colonization of host plants by their producers are toxic to animals; these metabolites are virulence or pathogenicity factors and mycotoxins at the same time. The most studied example is the trichothecene mycotoxin deoxynivalenol, which is produced by pathogenic *Fusarium* spp. infecting flowers and seeds of cereal crops and maize. Deoxynivalenol facilitates colonization of floral tissue, which leads to diseases designated *Fusarium* head blight in small-grain cereals and ear rot in maize. The same pathogens also colonize the roots and lower stems of cereal plants, causing seeding blight and root rot. Trichothecenes inhibit protein synthesis in eukaryotic cells,[1–3] which leads to pronounced toxic effects on animals.[3,4] Among the toxic effects of deoxynivalenol that mainly impact proliferating tissues, suppression of the immune system has been of most recent concern.[5,6] Accumulation of deoxynivalenol in cereal grains is an issue for food as well as feed safety, while accumulation in green plants and fodder crops affects only farm animals. In years with weather conditions favoring *Fusarium* infection, contamination of feed with deoxynivalenol causes serious problems in animal production. The maximum amount of deoxynivalenol allowed in food is limited by law in the European Union (EU) and many non-EU countries.

Mycotoxins relevant for food safety that act as virulence factors are rather rare. Apart from deoxynivalenol and its derivatives, the most prominent examples are the mycotoxins fusaric acid, enniatins, brefeldin, oxalate and phomopsin, each of which has been shown to act as a virulence factor in certain pathosystems. The role of the fumonisin mycotoxins in plant colonization has been supported by some data, but disputed by others. Certain toxicologically relevant mycotoxins, such as aflatoxins, ochratoxin A and zearalenone, do not seem to be involved in the etiology of any plant disease.

A specific group of fungal toxins involved in plant diseases are host-specific toxins. As with all other pathotoxins, these fungal metabolites modulate the physiology of their hosts, preventing defense responses and facilitating fungal colonization. The distinguishing feature of host-specific toxins is the limitation of their effect to certain varieties of susceptible crop species. Host-specific toxins are pathogenicity factors because they are necessary for the disease to occur. The first host-specific toxin was described in 1933 in Japan,[7] followed by a series of landmark studies on the toxins of *Alternaria* spp. These results were not adequately reflected by Western plant pathologists; it took 14 years until the first host-specific toxin discovered in the USA, named victorin, was described by graduate student Francis Meehan and her supervisor at United States Department of Agriculture (USDA).[8] After Robert Scheffer established the concept of host-specific toxins as a mainstream paradigm,[9] research on host-specific toxins attracted a large community of plant pathologists, but progress was slow due to the low performance of the analytical methods available in the 1940s to the 1960s for the detection of fungal toxins in plant tissue, as well as the limited options for structure elucidation. For instance, the constituents of victorin were characterized[10] and their order was established 38 years after the discovery of the toxin.[11]

Another host-specific toxin that was studied in great detail was HC toxin, named after *Helminthosporim carbonum*, which is an old name for *C. carbonum*. In 1938, a new devastating leaf spot disease of maize caused by *Cochliobolus carbonum* was found in Iowa on the maize variety Proudfit Reid. The diseases spread like wildfire over the continent, but was soon curbed by the identification of resistant varieties. Phytopathologists needed 27 years to find out that a host-specific toxin played a key role in the spread of the disease,[12] and further 17 years to elucidate its structure. Interestingly, several laboratories working on the structure of HC toxins accomplished the task and published the structure at the same time,[13–16] though one of them had to correct the configuration of one of the alanine residues.[17] The resolve of plant pathologists before 1990 to continue studying host-specific toxins over the decades in spite of the disappearance of most diseases in which these toxins played a role due to the introduction of resistant cultivars is commendable (Figure 8.2). Thanks to this dedication, plant pathology provided an indispensable contribution to the biological chemistry of fungi. This advancement would not have been possible without enormous field trials that were inadvertently carried out by farmers around the world each year by growing millions of plants in genetically homogenous monocultures. The immense selection pressures they maintain on a large scale make rare events visible, such as the horizontal transfer of an entire biosynthetic pathway for a host-specific toxin from one fungus to another. In 1969, many farmers in the Midwest lost their entire harvest due to Southern corn leaf blight; they might derive some satisfaction from having contributed to the genesis of a pathosystem with fascinating features such as maternally inherited susceptibility of the host and aberrant recombination at a toxin biosynthesis locus of the pathogen. This system would never be found in wild nature.

Most host-specific toxin studies so far have focused on polyketides and non-ribosomal peptides. The characteristic feature of host-specific toxins is high toxicity to susceptible cultivars with LD_{50} values orders of magnitude lower than for resistant cultivars and non-host plant species. Pathogen races producing host-specific toxins may co-exist with races not producing the

Timeline of the research on HC toxin

- 1938: Outbreak of Northern leaf spot
- 1965: HC toxin discovered
- 1982: Structure of HC toxin elucidated
- 1992: Detoxification in resistant plants detected
- 1997: Loss of detoxification in susceptible cultivars explained
- 2002: Protection against HC toxin by its producer clarified

Figure 8.2 Timeline of the research on host-specific toxins.

toxin without being noticed; introduction of a susceptible variety leads to a dramatic spread of the toxin-producing race, which may cover an entire continent within a few years. Highly specific interactions of host-specific toxins with molecular targets in host cells resembles incompatible interactions in pathosystems underlying the gene-for-gene hypothesis, but the molecular recognition that occurs between host-specific pathotoxins and their targets in the host cell leads to susceptibility rather than resistance. Similarly to the loss of resistance caused by mutations of avirulence factors or their receptors in plant cells,[18,19] the effects of host-specific toxins on host plants can easily be disrupted by modifying their targets. Mutations of toxin targets and the detoxification activities of plants account for the existence of resistant cultivars and accessions in all pathosystems involving host-specific toxins. Breeding crops for resistance to pathogens depending on host-specific toxins is therefore fast and efficient; diseases involving host-specific toxins are rarely encountered in plant production today. In the past, however, most cultivars of a particular crop grown in large areas were derived from a common genotype. Epidemics with disastrous consequences for plant production developed when the dominating genotype turned out to be susceptible to a particular host-specific toxin. Such epidemics wiped out the entire maize production on the North American continent within a few years, but the diseases disappeared after breeders provided resistant varieties.

Are host-specific toxins mycotoxins? Based on their structures, modes of action and the limited data on animal toxicity that are available, most host-specific toxins are likely to fulfill the definition of mycotoxins. The concentrations of these toxins in plant tissue are sufficient to break defense responses in susceptible cultivars, but might be too low to pose a threat to consumers' health. Another reason not to worry about host-specific toxins in food is the character of plant disease that depends on them. Crop diseases associated with the action of host-specific toxins are devastating. Because essentially no crop is harvested from susceptible cultivars infected with a toxin producer and no noticeable infection of resistant cultivars occurs, the exposure of consumers to host-specific toxins is likely to be minimal. A special case is AAL toxin, which is a host-specific toxin of *Alternaria alternata* f.sp. *lycopersici* on the tomato named after its producer.[20] The structure of AAL toxin resembles the fumonisin mycotoxins produced by maize pathogens *Fusarium verticillioides* and *Fusarium proliferatum* and by related species of the *Gibberella fujikuroi* species complex (Figure 8.3). In spite of profound differences between the sources and biological functions of fumonisins and AAL toxin, not only are their chemical structures related, but also their modes of action appear to be the same.

These metabolites interfere with the ceramide pathway by competitive inhibition of ceramide synthase due to their similarity with sphingosine. In tomato plants, this effect blocks defense responses against fungal infection, rendering the plants susceptible to *A. alternata* f.sp. *lycopersici*. In animals, the inhibition of ceramide synthesis causes apoptosis and interferes with the function of neurons, leading to species-specific syndromes such as leukoencephalomalacia (blind staggers) in horses, pulmonary edema in

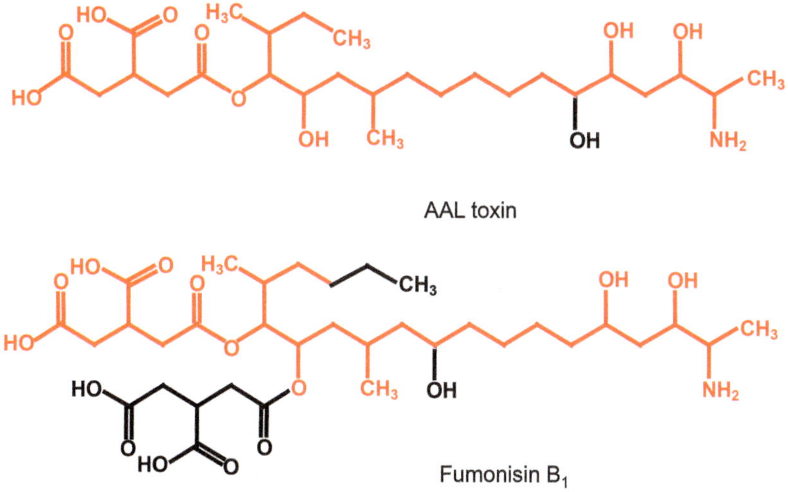

AAL toxin

Fumonisin B$_1$

Figure 8.3 Structures of the mycotoxin fumonisin B$_1$ produced by *Fusarium verticillioides* in maize and the host-specific pathotoxin AAL produced by *Alternaria alternata* in tomatoes.

pigs, liver cancer in rodents and allegedly also cancer of the esophagus in humans. Synthetic analogs of AAL toxin and fumonisins exert similar physiological effects on plant and animal cells.[21] Moreover, AAL toxin exerts the same effect on animal cells in tissue culture as fumonisins.[22] According to these results, AAL toxin is a mycotoxin. An analysis of the content of AAL toxin in tomato juice and tomato pulp on the market would show whether these products may pose a risk to consumers' health when tomato fruits that are used for their production are infected with *A. alternata* f.sp. *lycopersici*.

While the action of host-specific toxins can only be observed in specific combinations of a race (pathotype) of the fungal pathogen and a variety of the host plant, Ernst Gäumann and his co-workers and students in the 1950s postulated a central role for toxins in all plant diseases. In spite of efforts that lasted for a generation, they were unable to prove their hypothesis for fusaric acid and lycomarasmine, which they selected as model toxins. One reason for this failure was the same lack of sensitive analytical methods and spectroscopic tools for structure elucidation that hampered early research on host-specific toxins. The second reason, however, was even more serious, leading to *de facto* abandonment of Gäumann's hypothesis by future generations of plant pathologists: until the advent of molecular genetics, the support for Gäumann's hypothesis was limited to indirect evidence and correlations. Techniques for the selective inactivation of biosynthetic pathways in fungal pathogens became available three decades later. The use of indirect evidence in support of the role of toxins in plant diseases sparked sharp criticism (Figure 8.4).

Robert P. Scheffer's rediscovery of host-specific toxins in the 1960s helped keep interest in fungal toxins alive for two more decades,[23] before it was

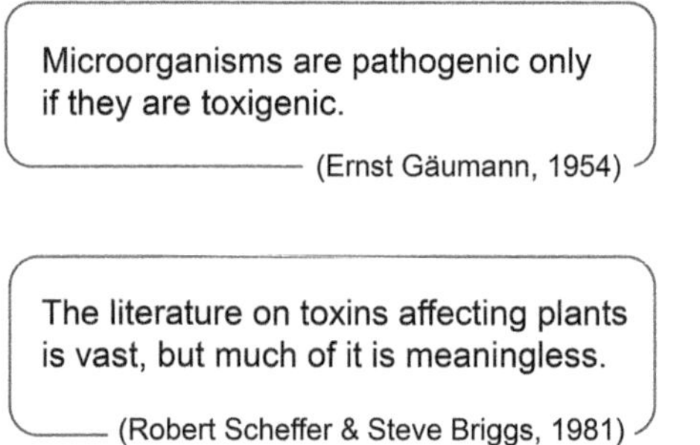

Figure 8.4 Conflicting views on the roles of toxins in plant diseases.

displaced by new fashions or paradigms. Mainstream research in plant pathology focused successively on extracellular enzymes, the gene-for-gene hypothesis and avirulence genes, signaling pathways in plant defense, molecular pathogenicity-associated patterns and, most recently, effectors. These waves of interest took most researchers and funds away from fungal toxins. A notable exception was the advancement of our understanding of the role of trichothecenes in pathogenicity when this was demonstrated for diacetoxyscirpenol[24] and deoxynivalenol.[25] Recent progress in fungal genomics and non-targeted metabolic profiling has sped up the assignment of chemical structures to biosynthetic pathways and triggered discoveries of new secondary metabolites. Some of these metabolites will likely turn out to be toxic to animals; they may eventually extend our list of compounds to be monitored and regulated in foods and feeds. While assessing the relevance of new fungal metabolites for food safety, one has to keep in mind that toxicity of fungal products to humans is merely a side effect of the ecological functions of these metabolites as agents of interference competition,[26] defense against mycoparasites,[27] defense against fungivorous invertebrates[28,29] and suppression of the defense response of host plants.[24,25,30–32]

Technical developments in metabolomics and fungal genomics provided plant pathologists with tools to readdress the role of fungal toxins in plant diseases. Gäumann's hypothesis in the form endorsed half a century ago was overstated; recent discoveries of numerous biosynthetic pathways for as-yet unknown secondary metabolites in genomes of fungal pathogens and the observation that genomes of plant pathogenic fungi harbor larger numbers of pathways for putative secondary metabolites as compared to saprophytes[33] revived Gäumann's ideas and vindicated his foresight.

8.2 Detoxification of Fungal Toxins in Plant Defense Against Fungal Infection

A key strategy helping to prevent yield and quality losses in plant production caused by fungal diseases is the cultivation of resistant varieties. In spite of tremendous progress that has been achieved in resistance breeding, complete and stable resistance against fungal pathogens has rarely been obtained. Dominant resistance encoded by single genes exists and is used extensively in breeding, but this is not available for pathogens producing toxicologically relevant mycotoxins. With rare exceptions, this kind of resistance occurs in pathosystems characterized by gene-for-gene relationships,[18] such as in powdery mildew of wheat, bean rust of beans and late blight of potatoes. Resistance responses in these systems is triggered by recognition of molecular signals produced by avirulence genes of the pathogens. The resistance is lost when the synthesis of the avirulence product is disrupted, its structure is changed or the matching resistance gene in the host plant is lost. Most pathogens that produce toxicologically relevant mycotoxins do not interact with their hosts according to gene-for-gene concept. Rather than escaping recognition, these pathogens thwart the defense responses of their hosts by killing cells at the infection site and extracting nutrients from the dead tissue. This lifestyle of plant pathogens is called necrotrophy. Breeding for resistance against necrotrophic pathogens relies on biochemically diverse polygenic traits known as horizontal resistance, which are not specific for a particular race or population of the pathogen. Horizontal resistance is less efficient than resistance based on the recognition of avirulence products, but it is more durable due to the involvement of multiple genes. Enzymatic detoxification of host-unspecific toxins is one of the mechanisms of horizontal resistance. Such detoxification is likely to convey very durable resistance, which the pathogen can only overcome by modifying the biosynthetic pathway in such a way that a toxin derivative is produced that is resistant to detoxification while retaining its toxicity to the host. Gradual increase of resistance of crops against mycotoxin-producing pathogens has been achieved over decades. With the exception of deoxynivalenol-producing pathogens (see below), it is unclear to what extent the detoxification of pathotoxins was involved in this progress.

By chance, the first plant resistance gene characterized at a molecular level encoded detoxification activity towards a host-specific toxin. The crop was maize and the pathogen was *Cochliobolus carbonum* race 1, which produced non-ribosomal cyclic tetrapeptide, designated HC toxin. The disease caused by *C. carbonum* race 1, designated Northern leaf spot, first occurred in Iowa, USA, in 1938. In a few years, the pathogen had spread over the entire corn-growing area of North America. Because it killed susceptible hosts, the epidemics threatened the entirety of corn production before breeders identified maize genotypes with a single-gene dominant resistance. The resistance gene was designated *Hm1* according to taxonomic naming of the asexual form of the pathogen *Helminthosporium carbonum*. The resistance exhibited Mendelian inheritance and was rapidly integrated into all commercial cultivars of maize. The gene product detoxified HC toxin by reducing the oxo group of 2-amino-9,10-epoxy-8-oxo-decanoic acid, which is one of the four amino acids of this cyclic tetrapeptide (Scheme 8.1).

The *Hm1* gene was the first plant resistance gene to be cloned and characterized at molecular level.[34,35] This function does not match the

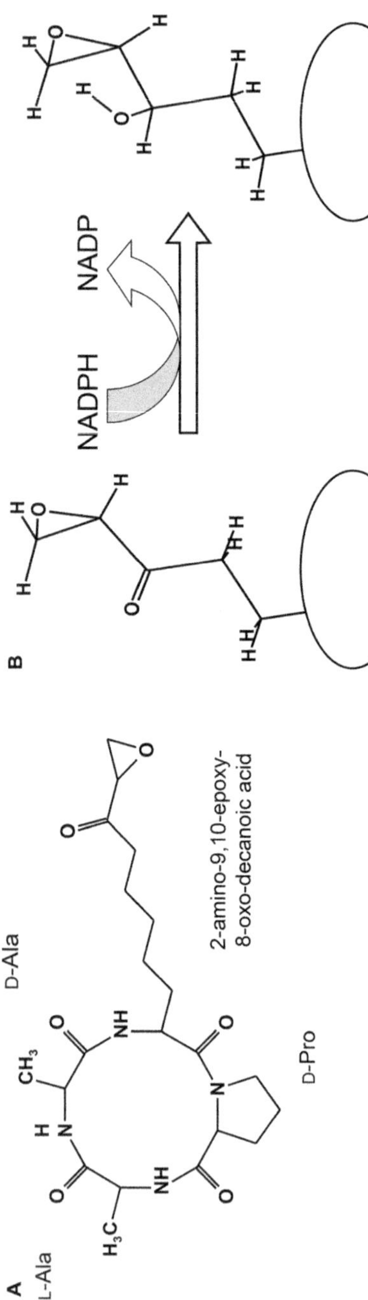

Scheme 8.1 Structure of HC toxin (A) and its detoxification in resistant plants (B).

mainstream paradigm, according to which plant resistance genes encode receptors for fungal elicitors that trigger signaling pathways, leading to defense responses. While old reviews on plant resistance genes included the *Hm1* gene,[36] this gene disappeared from more recent reviews. Some reviews excuse this by beginning with a statement that disease resistance is often determined by pairs of matching resistance gene products and their ligands or by limiting their treatment to recognition-dependent resistance, but many reviews silently ignore the existence of other kinds of resistance genes, presumably because the reviewers are not aware of them.[37–39] The bias towards fashionable topics seems more prominent in plant pathology than in other fields: the detoxification of many fungal pathotoxins in crops was demonstrated at a chemical level, but efforts to characterize the enzymes and genes involved have been limited.

Another instructive case of enzymatic detoxification of a host-specific pathotoxin by resistant plants was studied in the laboratory of M.S. Pedras at the University of Saskatchewan.[40] Destruxin B is a host-specific toxin produced by *Alternaria brassicae* that causes *Alternaria* blackspot disease of oilseed rape, mustard, broccoli and other crucifers. The cyclic depsipeptide destruxin B consist of four proteinogenic amino acids, β-alanine and 2-hydroxy-isohexanoic acid. Host plants transform the toxin into hydroxydestruxin B at a rate that is proportional to their resistance to infection with *A. brassicae*. Hydroxylated destruxin B was less phytotoxic than the parental toxin.[41] In the second detoxification step in resistant host plants, hydroxydestruxin B is glycosylated on the hydroxyl introduced by a plant enzyme[40] (Scheme 8.2).

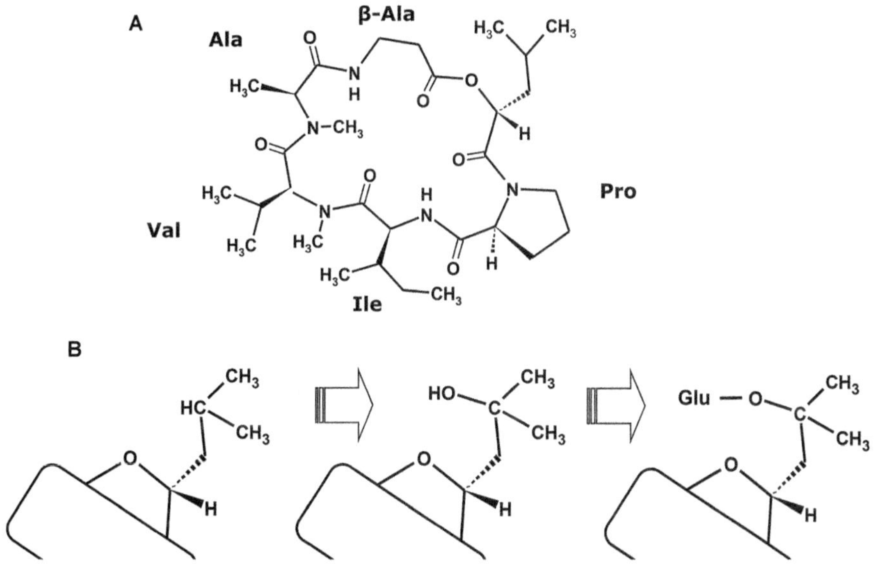

Scheme 8.2 Destruxin B and its detoxification in resistant plants. (A) Structure of cyclic despsipeptide destruxin B; (B) two-step transformation of destruxin B in leaves of *Sinapsis alba* (Pedras *et al.*[40]).

Interestingly, enzymatic modification of destruxin B by resistant plants occurs in the part of the molecule that distinguishes destruxin B from destruxin A, which is a virulence factor of the entomopathogenic fungus *Metarhizium anisopliae*.[42] Thus, three configurations of substituents on carbon 4 of 2-hydroxypentanoic acid on destruxins are known: a tertiary aliphatic carbon in 2-hydroxy-4-methylpentanoic acid of destruxin B; a tertiary alcohol in its detoxification product; and olefin on a secondary carbon in destruxin A. Roseotoxin B, which differs from destruxin B merely by methylation of proline[43] (Scheme 8.3), was found in *Trichothecium roseum*, which is a plant-pathogenic fungus; treatment of apple fruits with roseotoxin B indicated that the toxin might act as a virulence factor.[44] It would be interesting to see whether plants resistant to *T. roseum* detoxify reseotoxin B in the same manner as crucifers detoxify destruxin B.

M. anisopliae can be inoculated into living plants and colonize them without causing diseases symptoms, thus becoming an endophyte.[45] Inoculation of crop plants with *M. anisopliae* for the prevention of damage by herbivores raises food safety concerns because it may expose consumers to fungal toxins that have not occurred in food before. Biological control agents are advertised by proponents of organic farming as safe alternatives to chemical fungicides, and they are perceived as safe by most consumers, but all fungi used to control pathogens and pests produce mycotoxins that are considerably more toxic than any currently used pesticide.

Host-unspecific pathotoxins are virulence factor rather than pathogenicity factors. Although their detoxification does not stop infection, it enhances host plant resistance, alleviates diseases symptoms and reduces yield losses. Destruxin B is one of four host-unspecific pathotoxins that have been studied in great detail. The other three toxins are fusaric acid, deoxynivalenol and oxalate. Each of them is produced by a different pathogen. Fusaric acid and deoxynivalenol are detoxified by enzymatic activities that are naturally encoded in the genomes of the plant hosts of their producers. Detoxification of oxalate has only been observed in genetically engineered plants expressing suitable genes from other plants and fungi, though many plants natively produce proteins that are capable of oxalate detoxification (*e.g.* germin[46]).

Fusaric acid is one of the earliest mycotoxins to be discovered. It was described 80 years ago in a search for the causative agent of the pathologic elongation of rice plants afflicted by bakanae disease, which is caused by *Fusarium fujikuroi*.[47] The acute toxicity of fusaric acid is low; therefore, neither legal nor advisory limits for fusaric in food exist. Researchers argued that the occurrence of fusaric acid in feeds is relevant because it synergistically enhances the toxic effects of other mycotoxins,[48,49] but current legislature does not take interactions among food and feed contaminants into account. Fusaric acid is one of the oldest virulence factors studied in phytopathology. The first indications for the involvement of fusaric acid in plant diseases were reported by E. Gäumann for tomatoes.[50,51] In the following decades, further support for the role of fusaric acid in diseases

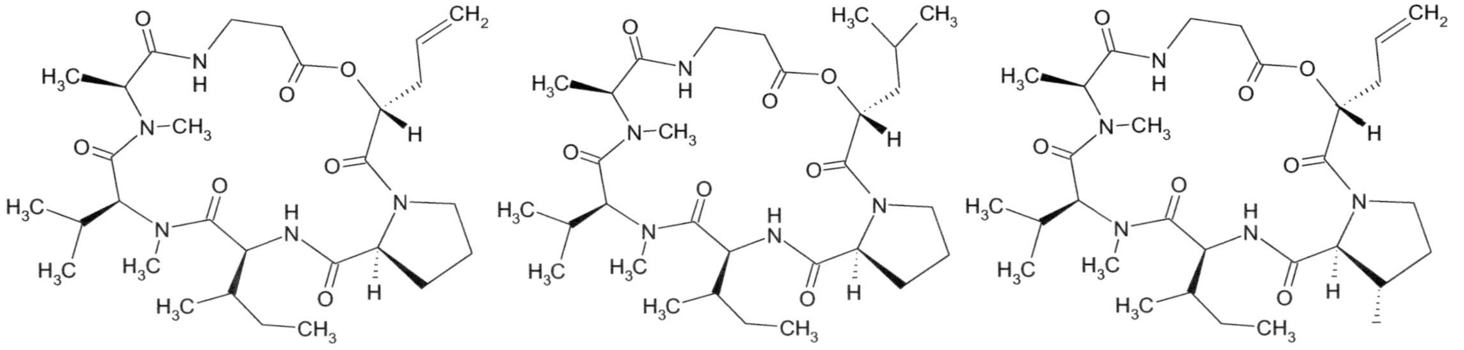

Destruxin A

Destruxin B

Roseotoxin B

Scheme 8.3 Structures of destruxin A produced by entomopathogenic *Metarhizium anisopliae*, destruxin B produced by pathogens of crucifers *Alternaria brassicae* and roseotoxin B produced by *Trichothecium roseum*.

Chapter 8

caused by *Fusarium* spp. was provided in tomatoes,[52] date palms,[53] *Orobanche*[54] and recently barley[55] and peas.[56] Direct tests of the function of fusaric acid in virulence were impossible before the biosynthetic pathway for the toxin was identified, which was recently achieved.[57] Gäumann's school established the detoxification of fusaric acid by resistant plants and demonstrated that the detoxification activity correlated with plant resistance to infection by pathogens producing fusaric acid. Such correlations were observed among host plant species as well as among varieties of the same crop. A remarkable accomplishment was the determination of the structure of the major detoxification product by Dieter Klüpfel, then a PhD student of Gäumann. The entire structural analysis was based on the chromatographic behavior of detoxification products of radioactively labeled fusaric acid and nine synthetic derivatives selected based on an educated guess.[58] Klüpfel exclusively used paper chromatography, but employed one- and two-dimensional separation with several elution systems. His indirect structure assignment preceded spectroscopic structure elucidation of fungal metabolites extracted from plants by decades. The major detoxification product of fusaric acid in tomato plants according to Klüpfel was an *N*-methylated amide of fusaric acid (Scheme 8.4).

Fifty years have passed since Klüpfel published his elaborate study of fusaric acid transformation by plants. It would be interesting and useful to verify his structural elucidation using current techniques to find out whether different plants possess the same transformation pathway and how common the activity is among hosts of pathogens producing fusaric acid.

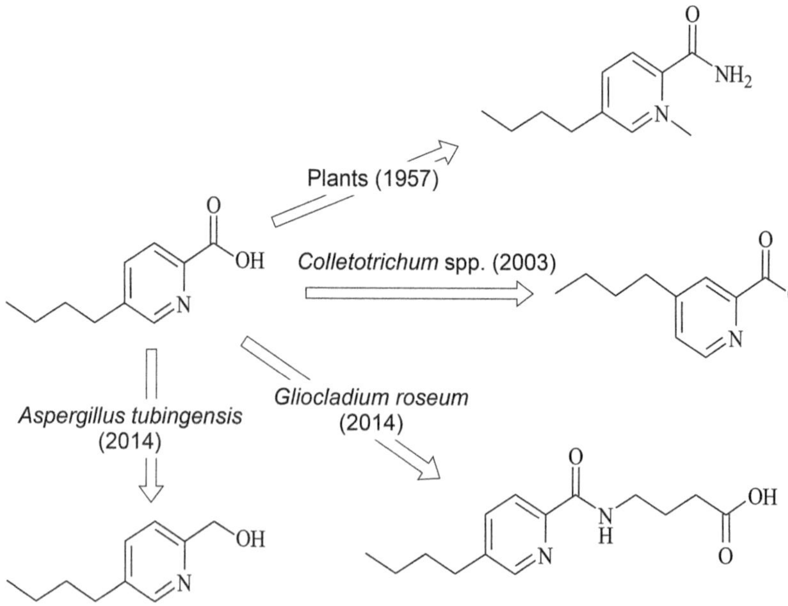

Scheme 8.4 Biological transformation of fusaric acid.

The straightforward next step will be the identification of the gene(s) responsible for the process as a basis for testing the hypothesis that the transformation of fusaric acid leads to detoxification that contributes to the resistance of plants to fungal infection. The techniques needed to answer these questions are now generally available. Recent research into the biotransformation of fusaric acid focused on microbial activities. With regards to the growing importance of pathogens producing fusaric acid, such as *Fusarium oxysporum* and *F. verticillioides*, it is now high time, half a century after Klüpfel's work, to relaunch research on the detoxification of fusaric acid in crops and its role in disease resistance.

The mycotoxin that has been studied most extensively as a virulence factor in plants is deoxynivalenol. Detoxification of deoxynivalenol by host plants was predicted to occur in maize tissue 30 years ago.[59] The detoxification products were detected by the same authors[60] who incubated ^{14}C-labeled deoxynivalenol with a cell culture of the wheat variety Frontana, which is highly resistant to *Fusarium* head blight, and analyzed the radioactively labeled products. Their results indicated that glycosylation of deoxynivalenol occurred, but the data were not sufficient for conclusive chemical proof. Similar experiments with maize cell cultures a few years later led to the unequivocal assignment of the transformation product to deoxynivalenol-3-glucoside.[61] In 2003, the formation of deoxynivalenol glucoside by an intact plant was shown for the first time, and the gene encoding for the deoxynivalenol-glycosylating enzymes responsible for the transformation (UDP-glucosyltransferase and UDP-glucose-deoxynivalenol transglucosylase) was characterized.[62] It came as no surprise that the plant upon which the study was carried out was *Arabidopsis thaliana*. The identification of a homologous glycosidase gene in wheat turned out to be difficult because of a large number of UDP-glucosyltransferases in the wheat genome; a promising candidate was identified in barley first.[63] Further information on the glycosylation of trichothecenes by plants can be found in a recent review.[64] A number of further detoxification reactions for deoxynivalenol have been described (Figure 8.5).

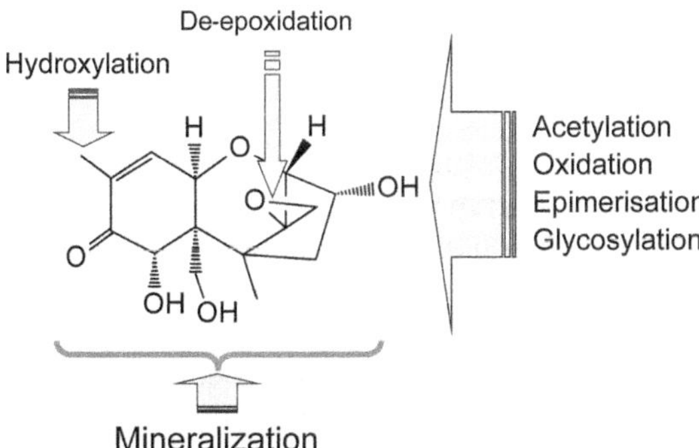

Figure 8.5 Detoxification of deoxynivalenol.

Research on the trichothecenes produced by *Fusarium* species colonizing cereal crops continues to generate fascinating results. The most recent discovery provides a rare example of modification of the structure of a pathotoxin.[65] A particular population of *F. graminearum* was found to produce deoxynivalenol derivatives that lack the oxo group on C8, which is the distinguishing feature of type B trichothecenes. The new trichothecenes were labeled NX-2 and NX-3. Although increased aggressiveness of *F. graminearum* due to this modification of deoxynivalenol structure has not been demonstrated, it is conceivable that a selection pressure exerted by certain wheat varieties may promote the spread of the new chemotype in wheat-growing regions. The implications for food safety are significant because the new trichothecene escapes monitoring by common high-performance liquid chromatography–tandem mass spectrometry protocols. It is likely that some of the *F. graminearum* strains reported not to produce deoxynivalenol or nivalenol in the past belong to this new chemotype. Furthermore, correlations between the biomass of *F. graminearum* in wheat ears and trichothecene levels might improve after the analysis of the predominant NX-3 trichothecene is integrated into standard multi-mycotoxin monitoring protocols.

8.3   Fungal Endophytes Producing Toxins

Certain fungi colonize living plants without causing visible diseases symptoms. Microorganisms colonizing plant tissue without harming the host are designated endophytes. They are often desirable in agricultural ecosystems because they improve plant growth and provide protection against pests.[66] Fungal endophytes may produce toxins that threaten the health of farm animals fed by the colonized plants and/or consumer health when plant organs containing the toxin are processed into food products. Tremorogenic mycotoxins produced by fungi colonizing forage grasses have been the most publicized mycotoxins of this kind because of the conspicuous symptoms shown by intoxicated livestock and their economic significance.

We have witnessed a boom in research into fungal endophytes and their deliberate inoculation in crops. Certain entomopathogenic fungi such as *Beauveria bassiana*, which has been used as an agent of biological control of insect pests for decades, were found to be able to colonize living plants, protecting them against insect herbivores.[67,68] A growing line of observations corroborate the view that endophytic growth may actually be the natural lifestyle of many entomopathogenic fungi,[69,70] which explains why artificial inoculation of crop plants with entomopathogenic fungi has been so successful. Fungal metabolites that are toxic to insects, such as beauvericin, tenellin and bassianin,[71] are likely to be involved in the protection of host plants against pests. Many metabolites of entomopathogenic fungi that are toxic to insects are also toxic to mammalian cells in tissue culture; these potential mycotoxins have to be monitored in food products derived from crops inoculated with entomopathogens. The best-known toxins of this kind are depsipeptides beauvericin, which is produced by *B. bassiana*, and destruxin A, which is produced by *M. anisopliae*.

Apart from protecting their hosts against herbivores, some fungal endophytes also protect their hosts against fungal diseases.[72–74] An interesting metabolic interaction between an endophytic fungus and its host was described for endophytes of the genus *Acremonium* colonizing *Taxus baccata*.[75] The fungus produces the linear nonapeptide leucinostatin A, which is a potent inhibitor of ATPases and is a lead structure in the development of antitrypanosomal drugs.[76] Because the toxin inhibits certain mammalian cells in tissue culture at nanomolar concentrations, it is likely to meet the definition of a mycotoxin. In the host plant, leucinostatin A is glycosylated at two hydroxyl groups (Figure 8.6).

It was speculated that glycosylation of leucinostatin A by *T. baccata* converts a potential pathogen into an endophyte.[75] This assumption is in line with the hypothesis of balanced antagonism,[72] which postulates that relationships between fungi colonizing plants and their hosts span a continuum of damage severity to the host, with symbiosis and parasitism demarking the extremes. Detoxification of toxins produced by endophytes may protect the host from phytotoxic effects. The concept of balanced antagonism is attractive because many pathogens of economic relevance in plant production are known to occur as endophytes in other hosts, or even in the same host under particular conditions. The causes of transition between pathogenic and endophytic lifestyles, however, are not understood. Well-studied examples are *F. verticillioides* and *F. oxysporum*. Current paradigms in plant pathology, however, do not support the balanced antagonism hypothesis. The establishment of compatibility between pathogens and their hosts occurs at two levels: by active disruption of non-host resistance *via* effectors injected into host cells; and by avoidance of race-specific recognition after effectors became avirulence products. This framework offers no space for balanced antagonism as a basis for an endophytic lifestyle.

Toxins of fungal endophytes are instrumental in conveying benefits to their host by protecting them against pests. These beneficial effects turn undesirable when the toxins jeopardize the health of livestock. Detoxification of these toxins by host plants is unlikely to occur spontaneously, but developing transgenic varieties possessing suitable detoxification activities is conceivable, with the obvious drawback of impairing resistance to herbivores.

8.4 Genetic Engineering of Crops for Detoxification of Fungal Toxins

8.4.1 Concept

A theoretical basis for resistance enhancement by detoxification of pathotoxins was established by Gäumann and his students in the 1950s. Because this happened decades before plant pathologists could even dream of engineering properties of crop plants, Gäumann and his disciples cannot be

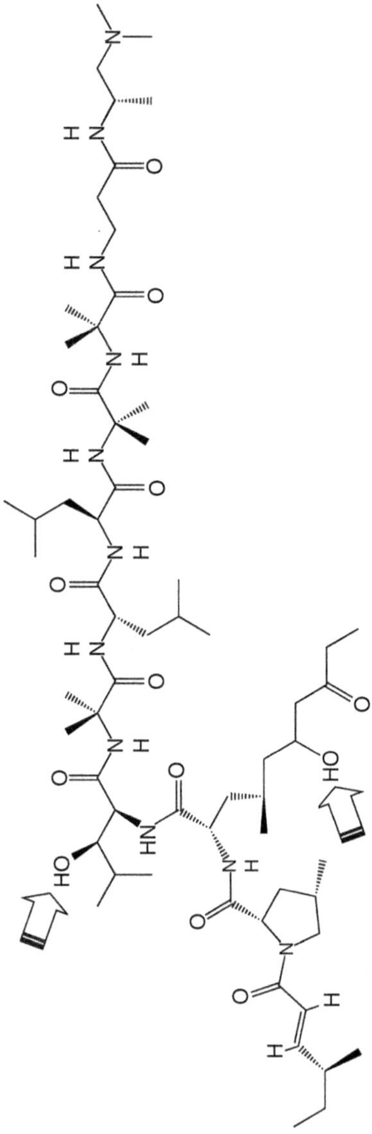

Figure 8.6 Glucosylation of leucinostatin A by *Taxus baccata*.

blamed for not having foreseen the applications of their concept in resistance engineering.

Even after genetic transformation of plants had been established, the potential of enzymes detoxifying fungal toxins for practical applications was not instantly recognized. The discovery of the degradation of the mycotoxin zearalenone by the mycoparasite *Gliocladium roseum* by El-Sharkawy and Abul-Hajj[77] provides a striking demonstration of this circumstance. Although the toxicological relevance of zearalenone ingested with food and feeds had been firmly established more than two decades earlier,[78–80] the authors did not suggest any application of their finding for the reduction of zearalenone contamination. The goal of their project was to identify microbial strains simulating mammalian metabolic processes. The only practical aspect of zearalenone detoxification discussed in their paper was its relevance for use as a growth-promoting agent for farm animals. If mammals possessed such an activity (which as we now know they do not), zearalenone-based growth promotants would be ineffective. The discovery of the biological degradation of zearalenone by El-Sharkawy and Abul-Hajj, which is now regarded as a seminal achievement in the field, was hardly noticed by the research community in the 1990s. When researchers at RIKEN (Saitama, Japan) launched a project on the biodegradation of zearalenone, instead of verifying El-Sharkawy and Abul-Hajj's work, they initiated an extensive *de novo* screening of microbial isolates obtained from soil and plants. As a result, they rediscovered the activity described by El-Sharkawy and Abul-Hajj (see details below).

Exactly at the time when the transformation of zearalenone by fungi was studied in Minnesota as a model for mammalian metabolism, Hideyoshi Toyoda and his colleagues in Kinki University in Osaka began developing a new concept that later became a paradigm for transgenic strategies against fungal toxins. They worked with fusaric acid, bringing back to Japan a toxin that was discovered there but studied by plant pathologists exclusively in Europe. (Research on fusaric acid in Japan focused on its potential for the treatment of high blood pressure, drug addiction and other human diseases.) Toyoda's team in the Laboratory of Plant Pathology of Kinki University first envisioned the application of intact toxin-degrading microbes as a means of preventing fungal diseases, but by as early as in 1988, they cloned the first fungal gene that protected bacteria from fusaric acid and foresaw the use of such genes in transgenic crops. In spite of not having implemented their idea into practice, the group is given credit for the concept of detoxification of fungal toxins in transgenic plants. This is one among many topics in plant–microbe interactions in which researchers in Japan radically advanced the field.

8.4.2 Sources of Enzymes and Genes for Detoxification of Fungal Toxins

Any organism possessing a suitable activity can serve as a source of genes for the detoxification of mycotoxins. Crop plants possess activities that detoxify some pathogenicity and virulence factors. Such activities were selected for during the co-evolution of plants with their pathogens and are present in crop varieties and related wild species. The most studied detoxification activities of plants target the host-specific toxins HC toxin and destruxin B and the host-unspecific toxins fusaric acid, trichothecenes and oxalate. The enzymatic activities of a plant might, by chance, detoxify mycotoxins that do not play any role in plant disease; glycosylation of zearalenone in maize[81] and barley roots[82] appear to be results of such accidental matches between the specificities of UDP-glucosyltransferases and the resorcylic acid moiety of zearalenone.

Detoxification activities that do not belong to the enzymatic repertoire of plants can, with a low likelihood, be selected *in vitro* by treating cell cultures with toxins. Plant breeders often selected tissue cultures for resistance to fungal toxins in the expectation that plants regenerated from toxin-resistant calli would be resistant to toxin-producing fungi. How successful the strategy was is unclear. Regenerated plants resistant to fungal infection were obtained in some laboratories, but untreated controls were rarely subjected to regeneration and testing for resistance to infection in a comparable effort, raising a question as to whether enhanced resistance was not due to the selection of somaclonal variants.

The mechanism of enhanced resistance of regenerated plant cultures treated with fungal toxins was rarely studied in outcome-oriented resistance breeding. An exception was a project on the selection of resistance to brefeldin in safflower (*Carthamus tinctorius* L.), carried out in the laboratory of Ulrich Matern in the University of Freiburg. Brefeldin is a pathogenicity factor of *Arternaria carthami* Chowdhury, which is the causal agent of a devastating *Alternaria* leaf blight disease of safflower. Brefeldin A is macrocyclic lactone and plants produce numerous esterases with a wide range of specificities. These two circumstances nourished a working hypothesis that spontaneous mutations arising in tissue cultures might extend the specificity of one of the esterases of safflower, enabling it to hydrolyze the lactone bond of brefeldin. In spite of repeated trials with different regimes of gradually increasing brefeldin concentration, no hydrolysis of the toxin was detected.[83] Disappointed by these frustrating results, one researcher went to the backyard of the building, brought in a soil sample, incubated it for 24 hours in minimal medium with brefeldin and plated the culture on agar plates amended with brefeldin. One of the strains isolated from the plates, identified as *Bacillus subtilis*, degraded brefeldin A in pure culture by hydrolysis of its lactone bond (Scheme 8.5). The strain, which was labeled *B. subtilis* BG3, degraded hydrolyzed brefeldin A

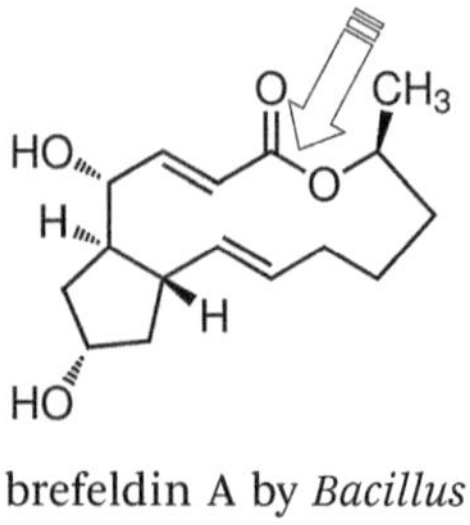

Scheme 8.5   Detoxification of brefeldin A by *Bacillus subtilis*.

further, but the products were not characterized.[84] The gene encoding brefeldin lactonase was cloned, but the team disbanded before transgenic safflower plants could be created. The experience with brefeldin A corroborates a common view that the chances of finding activities that degrade a particular organic compound are much better in microbes than in plants.

While plants rarely encounter toxins except when they are produced by pathogens, herbivores—especially the generalists among them—consume a plethora of plant toxins, most of which are believed to have evolved as protection against herbivores. Potent detoxification activities of herbivores towards toxins may therefore be expected. While such activities have been described, they do not match the diversity, efficiency and widespread occurrence of degradation activities of bacteria and fungi. It appears straightforward that microorganisms inhabiting environments where they frequently experience exposure to a particular mycotoxin are likely to develop matching biotransformation activities. Such a transformation is regarded as detoxification when the toxin acts as an antibiotic, inhibiting the growth of the microorganism. Other degradation activities might have been selected in evolution for their catabolic function as the first steps of pathways releasing metabolically usable energy.

8.4.3 Examples of Detoxification Activities

8.4.3.1 *Zearalenone*

As mentioned in the introduction to this section, detoxification of zearalenone by *G. roseum* was described in 1988 in the University of Minnesota[77] and rediscovered in 2002 in RIKEN, Japan.[85] The active strain identified as *Clonostachys rosea* IFO 7063 turned out to be nearly isogenic with *G. roseum* NRRL1829, identified as a zearalenone detoxifier strain in Minnesota 14 years ago. In the follow-up work, the RIKEN team purified and characterized the enzyme, sequenced fragments of the protein, design degenerated PCR primers based on peptide sequences and eventually cloned the gene encoding zearalenone lactonase, which they named *zhd101*.[86] At the same time, DuPont/Pioneer Hi-Bred in the USA pursued the cloning of the gene from *G. roseum* NRRL1829 based on differential transcript profiling. Their strategy was based on the observation that the enzyme's synthesis is induced by zearalenone and α-zearalenol, but not by β-zearalenol, though all three compounds are good substrates for lactonase activity.[87] DuPont's group cloned the gene, named it *zes2* and protected its use in microorganisms, mono- and di-cotyledonous plants and non-human animal cells by patents.[88,89] In 2003, RIKEN filed for patent protection for the same gene designated *zhd101* with World Intellectual Property Organization (WIPO),[104] but their application has not entered the European phase, apparently due to prior art cited by the examiner.

According to El-Sharkawy and Abul-Hajj,[77] degradation of zearalenone by *G. roseum* proceeds in three steps (Scheme 8.6): (i) the lactone bond is

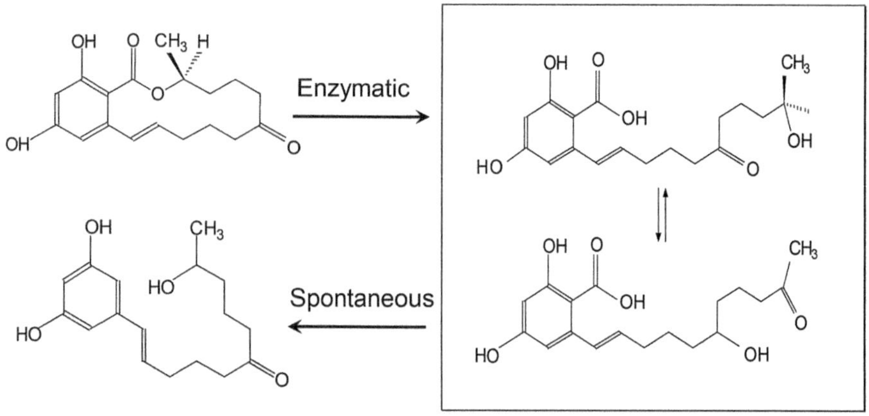

Scheme 8.6 Detoxification of zearalenone by *Clonostachys rosea.*

enzymatically hydrolyzed; (ii) the product is spontaneously decarboxylated; and (iii) the resulting hydroxyketone 1-(3,5-dihydroxyphenyl)-10′-hydroxy-1′-undecene-6′-one undergoes isomerization into 1-(3,5-dihydroxyphenyl)-6′-hydroxy-1′-undecene-10′-one. According to the authors, the hydroxyketone engaged in a rapid isomerization at room temperature; chromatography at $-20\,^{\circ}\mathrm{C}$ was necessary to separate the isomers for spectroscopic analysis. The authors hypothesized that internal hydride transfer through a six-membered transition state was responsible for the isomerization, and backed up their hypothesis by referring to the analysis of alkaline hydrolysis of zearalenone,[90] which led to identical products (Scheme 8.6).

The RIKEN team, working with the same enzyme and substrate, obtained 1-(3,5-dihydroxyphenyl)-10′-hydroxy-1′-undecene-6′-one, labeling it **2**, as the only product. In contrast to El-Sharkawy and Abul-Hajj,[77] they have not observed any isomerization. To back up their findings, they wrote: "The fact that the base treatment of **1** (zearalenone) with a 10% sodium bicarbonate solution under reflux also gave **2** supports the mechanism for this biotransformation by *Clonostachys rosea* IFO 7063," referring to the original description of zearalenone[90] cited by El-Sharkawy and Abul-Hajj[77] in support of isomerization. An inspection of this paper reveals that alkaline hydrolysis of zearalenone was followed by decarboxylation and isomerization of the same kind as described by El Sharkawy and Abul-Hajj.[77] The publication therefore does not back up the exclusive production of 1-(3,5-dihydroxyphenyl)-10′-hydroxy-1′-undecene-6′-one described in RIKEN. NMR data published by RIKEN showed the same signals for C11′ as reported by El-Sharkawy and Abul-Hajj for 10′-hydroxy-6′-ketone; signals for 6′-hydroxy-10′-ketone, which should be present in a mixture of both compounds, were missing. The controversy thus remains unresolved.

Why does *C. rosea* degrade zearalenone? The fungus is a mycoparasite that is presumably exposed to zearalenone while preying on zearalenone producers. In their description of the zearalenone lactonase gene, the RIKEN team wrote that lactonase activity in *C. rosea* was "enhanced by the addition of an oestrogenic compound ZEN without antifungal activity".[86] The inducibility of zes2/*zhd101* expression by zearalenone and α-zearalenol was described before,[87] but the antifungal effects of zearalenone were discovered 5 years later.[27] These effects inspired the hypothesis that zearalenone lactonase protected *C. rosea* from zearalenone produced as a defense metabolite by its prey, which was corroborated by showing that inactivation of the gene encoding zearalenone lactonase increased the susceptibility of *C. rosea* to zearalenone.[27] Further support for the hypothesis came from a study of the interaction of a gene disruption mutant of *C. rosea* with the zearalenone producer *F. graminearum*.[91] In addition to *C. rosea*, other mycoparasites apparently possess a gene for zearalenone lactonase.[92]

The basidiomycete yeast *Trichosporon mycotoxinivorans*, isolated from the termite *Mastotermes darwiniensis*, degrades zearalenone by a different mechanism. The macrocyclic ring is opened at the ketone group C6′, generating 5-([2,4-dihydroxy-6-{5-hydroxypent-1-en-1-yl}benzoyl]oxy)hexanoic acid as a product.[93] The authors hypothesized that oxygen was inserted into the macrocyclic ring next to the oxo group before the ring is opened, resembling Baeyer–Villiger oxidation of ketones. Because of the complex mechanism involved, the pathway appears less suitable for expression in other organisms than the degradation activity of *C. rosea*. Regarding the biological function of zearalenone degradation in *T. mycotoxinivorans*, we conjecture that zearalenone resembles the natural substrate of the enzyme(s) by chance rather than being the target for which the activity was selected.

8.4.3.2 Fusaric Acid

Fusaric acid was the first fungal toxin targeted by the development of detoxification strategies for transgenic plants, as described in the introductory part of this section. In the 1980s and 1990s, Hideyoshi Toyoda and his colleagues at Kinki University, Osaka, isolated a number of microorganisms detoxifying fusaric acid, cloned the genes and demonstrated that when expressed in heterologous hosts, these genes provided protection against damage inflicted by fusaric acid.[94] Although they have not expressed any of their genes in plants, they disclosed the idea in publications and in patents owned by Suntory, Ltd and Daikin Industries, Ltd.[95,96] For most of these activities, chemical detoxification was proven by chemical methods or by the loss of phytotoxicity after incubating fusaric acid with the microorganisms.[94,97,98] Interestingly, the structures of the products of fungal and bacterial activities for detoxifying fusaric acid that were discovered at Kinki University were not determined. Our laboratory is in the process of reproducing these results with the goal of clarifying the underlying chemistry. However, regarding a gene from *Pseudomonas cepacia* UK1, labeled 'fusaric acid-resistance gene', it remains unclear whether the strain detoxified fusaric acid enzymatically or merely transported it actively out of the cells.[99] The authors wrote that they isolated the strain on minimal medium containing only fusaric acid as a carbon source, implying that degradation must have been involved. A study of efflux pumps in *Pseudomonas aeruginosa* revealed that FusA, found in Japan to be responsible for fusaric acid resistance in *P. cepacia* UK1,[99] possessed homology to *oprK*, which encodes an efflux pump involved in resistance to antibiotics.[100] Independent sequence analysis by a group studying the involvement of membrane proteins in antibiotic resistance in *P. cepacia*, however, could not find any significant similarity between *opcM* and FusA.[101] When researchers at Kinki University published their results, it was not mandatory to deposit the sequences in GeneBank. The inferior quality of the scan of the article available from the website of the journal could have led to errors of optical character recognition, which may have contributed to this controversy.

Many laboratories have described transformations of fusaric acid by microbial strains since the pioneering work at Kinki University. An overview of the chemistry is shown in Scheme 8.4. Most unusual among these transformations was the transfer of *n*-butyl from position 5 to 4 of picolinic acid. The authors labeled the product isofusaric acid and suggested that the enzyme isofusaric acid synthase was involved.[102] Enzymatic transfer of butyl groups or aliphatic chains in general along aromatic rings is uncommon. The authors pointed out that isochorismate synthase catalysis is a similar process, but we do not think that this is an adequate analogy. Conversion of chorismate to isochorismate superficially resembles migration of a hydroxyl along a benzene ring, but a complex redox mechanism is involved; the incoming hydroxyl is derived from water rather than from an intramolecular transfer. Because isofusaric acid is much less phytotoxic to plants, further investigation of this transformation is attractive, and not only for academic reasons. Let us hope that the strain used in this work, which was labeled as a 'Colletotrichum spp.' in the publication and apparently not deposited in culture collection, is still alive and available for closer analysis. The conversion of fusaric acid to an *N*-methylated amide in plants was described in Gäumann's laboratory;[58] reduction to fusarinol by *Aspergillus tubingensis* was recently discovered by a consortium of three laboratories in the USA;[103] and conjugation of fusaric acid with γ-aminobutanoic acid carried out by another fungus was discovered by Yi Kuang in our laboratory (Yi Kuang *et al.*, in preparation).

8.5 Perspectives for Genetically Modified Crops Detoxifying Fungal Toxins

The idea of improving the capability of crop plants to counteract fungal toxins by enzymatic activities originating from other organisms is now 30 years old. The strategy relies on genetics without the introduction of chemicals or microorganisms into agricultural systems, which is attractive from economic and ecological points of view. In spite of this advantage, no implementation in plant production has been realized so far. One reason for this is that detoxification activities against most toxins that act as pathogenicity factors exist in crops. These activities have been used by plant breeders, often without knowing that the resistance traits they were introducing into new varieties relied on detoxification. But what about other fungal toxins accumulating in plants in the field? Potential targets are virulence factors and also fungal metabolites that do not harm plants but endanger our health. The technology promising to destroy these toxins in the plant tissue by highly specific enzymatic activities is mature. The sources of suitable enzymatic activities are known and many genes have been cloned and characterized. The obstacles to the implementation of this are of political and psychological in nature. Food companies rely on genetically engineered organisms, but they do not declare the origins of vitamins, enzymes and processing aids in drinks, pastries, cheeses and other products because they do not have to. Opponents of genetically modified (GM) crops are well aware of the situation, but in a peculiar alliance with the industry, they keep consumers in ignorance; admitting that GM products have been consumed by all of us for years without problem would destroy their agenda. Under these circumstances, no food company would use GM crops in their major products in Europe. Tropical countries suffer from fungal contamination and many of them do not have surpluses of cheap food; these markets are more likely to implement GM crops of different kinds, including crop varieties equipped for the detoxification of fungal toxins. It is impossible to predict the timeline for this implementation, but in the long run, even Europe cannot ignore the opportunity to remove poisonous natural products from food using biotechnology.

References

1. Y. Ueno, M. Nakajima, K. Sakai, K. Ishii, N. Sato and N. Shimada, Comparative toxicology of trichothec mycotoxins: inhibition of protein synthesis in animal cells, *J. Biochem.*, 1973, **74**, 285–296.
2. E. Cundliffe, M. Cannon and J. Davies, Mechanism of inhibition of eukaryotic protein synthesis by trichothecene fungal toxins, *Proc. Natl. Acad. Sci. U. S. A.*, 1974, **71**, 30–34.
3. B. A. Rotter, D. B. Prelusky and J. J. Pestka, Toxicology of deoxynivalenol (vomitoxin), *J. Toxicol. Environ. Health*, 1996, **48**, 1–34.
4. M. Maresca, From the gut to the brain: Journey and pathophysiological effects of the food-associated trichothecene mycotoxin deoxynivalenol, *Toxins*, 2013, **3**(5), 784–820.
5. J. J. Pestka, H.-R. Zhou, Y. Moon and Y. J. Chung, Cellular and molecular mechanisms for immune modulation by deoxynivalenol and other trichothecenes: unraveling a paradox, *Toxicol. Lett.*, 2004, **153**, 61–73.
6. B.-K. Choi, S.-H. Jeong, J.-H. Cho, H.-S. Shin, S.-W. Son, Y.-K. Yeo and H.-G. Kang, Effects of oral deoxynivalenol exposure on immune-related

parameters in lymphoid organs and serum of mice vaccinated with porcine parvovirus vaccine, *Mycotoxin Res.*, 2013, **29**, 185–192.

7. S. Tanaka, Studies on the black spot disease of the Japanese pears (*Pyrus serotina*), *Mem. College Agr. Kyoto Imp. Univ.*, 1933, **28**, 1–31.

8. F. Meehan and H. C. Murphy, Differential phytotoxicity of metabolic by-products of *Helminthosporium victoriae*, *Science*, 1947, **106**, 270–271.

9. R. P. Scheffer and R. S. Livingston, Host-selective toxins and their role in plant diseases, *Science*, 1984, **223**, 17–21.

10. V. Macko, T. J. Wolpert, W. Acklin, B. Jaun, J. Seibl, J. Meili and D. Arigoni, Characterization of victorin C, the major host-selective toxin from *Cochliobolus victoriae*: Structure of degradation products, *Experientia*, 1985, **41**, 1366–1370.

11. T. J. Wolpert, V. Macko, W. Acklin, B. Jaun, J. Seibl, J. Meili and D. Arigoni, Structure of victorin C, the major host-selective toxin from *Cochliobolus victoriae*, *Experientia*, 1985, **41**, 1524–1529.

12. R. P. Scheffer and A. J. Ullstrup, a host-specific toxic metabolite from helminthosporium carbonum, *Phytopathology*, 1965, **55**, 1037–1038.

13. J. M. Liesch, C. C. Sweeley, G. D. Staffeld, M. S. Anderson, D. J. Weber and R. P. Scheffer, Structure of HC-toxin, a cyclic tetrapeptide from *Helminthosporium carbonum*, *Tetrahedron*, 1982, **38**, 45–48.

14. M. L. Gross, D. McCrery, F. Crow, K. B. Tomer, M. R. Pope, L. M. Ciuffetti, H. W. Knoche, J. M. Daly and L. D. Dunkle, Tandem mass spectrometry for the structural determination of backbone-modified peptides, *Tetrahedron Lett.*, 1982, **23**, 5381–5384.

15. J. D. Walton, E. D. Earle and B. W. Gibson, The structure and conformation of HC-toxin, *Biochemical and Biophysical Research Communications*, 1982, **107**, 785–794.

16. M. R. Pope, L. M. Ciuffetti, H. W. Knoche, D. McCrery, J. M. Daly and L. D. Dunkle, *B* Structure of an amino acid analog of the host-specific toxin from *Helmintosporium carbonum*, *Biochemistry*, 1983, **22**, 3502–3506.

17. M. Kawai, D. Rich and J. Walton, The structure and conformation of HC-toxin, *Biochem. Biophys. Res. Commun.*, 1983, **111**, 398–403.

18. R. Laugé and P. J. G. M. De Wit, Fungal avirulence genes: structure and possible functions, *Fungal Genet. Biol.*, 1998, **24**, 285–297.

19. T. Boller and G. Felix, A renaissance of elicitors: perception of microbe-associated molecular patterns and danger signals by pattern-recognition receptors, *Annu. Rev. Plant Biol.*, 2009, **60**, 379–406.

20. D. Gilchrist, H. Wang and R. Bostock, Sphingosine related-mycotoxins in plant and animal diseases, *Can. J. Bot.-Rev. Can. Bot.*, 1995, **73**, 459–467.

21. H. K. Abbas, T. Tanaka and W. T. Shier, Biological activities of synthetic analogues of *Alternaria alternata* toxin (AAL-toxin) and fumonisin in plant and mammalian cell cultures, *Phytochemistry*, 1995, **40**, 1681–1689.

22. W. Wang, C. Jones, J. Ciacci-Zanella, T. Holt, D. G. Gilchrist and M. B. Dickman, Fumonisins and *Alternaria alternata lycopersici*

toxins: sphinganine analog mycotoxins induce apoptosis in monkey kidney cells, *Proc. Natl. Acad. Sci. U. S. A.*, 1996, **93**, 3461–3465.

23. R. B. Pringle and R. P. Scheffer, Host-specific plant toxins, *Annu. Rev. Phytopathol.*, 1964, **2**, 133–156.

24. A. Desjardins, T. Hohn and S. McCormick, Effect of gene disruption of trichodiene synthase on the virulence of *Gibberella pulicaris*, *Mol. Plant-Microbe Interact.*, 1992, **5**, 214–222.

25. R. Proctor, T. Hohn and S. McCormick, Reduced virulence of *Gibberella zeae* caused by disruption of a trichothecene toxin biosynthetic gene, *Mol. Plant-Microbe Interact.*, 1995, **8**, 593–601.

26. J. B. Gloer and S. Truckenbrod, Interference competition among coprophilous fungi: production of (+)isoepoxydon by *Poronia punctata*, *Appl. Environ. Microbiol.*, 1988, **54**, 861–864.

27. J. Utermark and P. Karlovsky, Role of zearalenone lactonase in protection of *Gliocladium roseum* from fungitoxic effects of the mycotoxin zearalenone, *Appl. Environ. Microbiol.*, 2007, **73**, 637–642.

28. M. Rohlfs, M. Albert, N. P. Keller and F. Kempken, Secondary metabolites protect mould from fungivory, *Biol. Lett*, 2007, **3**, 523–525.

29. K. Döll, S. Chatterjee, S. Scheu, P. Karlovsky and M. Rohlfs, Fungal metabolic plasticity and sexual development mediate induced resistance to arthropod fungivory, *Proc. R. Soc. B*, 2013, **280**, 20131219.

30. E. Gäumann, Bacterial phytotoxin: Mechanism of action, *Experientia*, 1954, **13**, 198–204.

31. M. Herrmann, R. Zocher and A. Haese, Effect of disruption of the enniatin synthetase gene on the virulence of *Fusarium avenaceum*, *Mol. Plant-Microbe Interact.*, 1996, **9**, 226–232.

32. N. Möbius and C. Hertweck, Fungal phytotoxins as mediators of virulence, *Curr. Opin. Plant Biol.*, 2009, **12**, 390–398.

33. S. Kroken, N. L. Glass, J. W. Taylor, O. C. Yoder and B. G. Turgeon, Phylogenomic analysis of type I polyketide synthase genes in pathogenic and saprobic ascomycetes, *Proc. Natl. Acad. Sci. U. S. A.*, 2003, **100**, 15670–15675.

34. G. S. Johal and S. Briggs, Reductase activity encoded by the HM1 disease resistance gene in maize, *Science*, 1992, **258**, 985–987.

35. R. Meeley, G. Johal, S. Briggs and J. Walton, Genetic patterns of plant host-parasite interactions, *Plant Cell*, 1992, **4**, 71–77.

36. F. L. W. Takken and M. H. A. J. Joosten, Plant resistance genes: their structure, function and evolution, *Eur. J. Plant Pathol.*, 2000, **106**, 699–713.

37. J. Ellis, P. Dodds and T. Pryor, Structure, function and evolution of plant disease resistance genes, *Curr. Opin. Plant Biol.*, 2000, **3**, 278–284.

38. J. M. McDowell and B. J. Woffenden, Plant disease resistance genes: recent insights and potential applications, *Trends Biotechnol.*, 2003, **21**, 178–183.

39. R. Zhang, F. Murat, C. Pont, T. Langin and J. Salse, Paleo-evolutionary plasticity of plant disease resistance genes, *BMC Genomics*, 2014, **15**, 187.

40. M. S. Pedras, I. L. Zaharia, Y. Gai, Y. Zhou and D. E. Ward, *In planta* sequential hydroxylation and glycosylation of a fungal phytotoxin: Avoiding cell death and overcoming the fungal invader, *Proc. Natl. Acad. Sci. U. S. A.*, 2001, **98**, 747–752.

41. M. S. Pedras, C. J. Biesenthal and I. L. Zaharia, Comparison of the phytotoxic activity of the phytotoxin destruxin B and four natural analogs, *Plant Sci.*, 2000, **156**, 185–192.

42. S. Pal, R. J. S. Leger and L. P. Wu, Fungal peptide destruxin a plays a specific role in suppressing the innate immune response in *Drosophila melanogaster*, *J. Biol. Chem.*, 2007, **282**, 8969–8977.

43. J. P. Springer, R. J. Cole, J. W. Dorner, R. H. Cox, J. L. Richard, C. L. Barnes and D. Van der Helm, The fungal spore and disease initiation in plants and animals, *J. Am. Chem. Soc.*, 1984, **106**, 2388–2392.

44. M. Zabka, K. Drastichová, A. Jegorov, J. Soukupová and L. Nedbal, Direct evidence of plant-pathogenic activity of fungal metabolites of *Trichothecium roseum* on apple, *Mycopathologia*, 2006, **162**, 65–68.

45. G. Hu and R. J. S. Leger, Field studies using a recombinant mycoinsecticide (*Metarhizium anisopliae*) reveal that it is rhizosphere competent, *Appl. Environ. Microbiol.*, 2002, **68**, 6383–6387.

46. B. G. Lane, Oxalate, germin, and the extracellular matrix of higher plants, *FASEB J.*, 1994, **8**, 294–301.

47. T. Yabuta, K. Kobe and T. Hayashi, Biochemical studies of the 'bakanae' fungus of rice. I. Fusarinic acid, a new product of the 'Bakanae' fungus, *Zeitschrifüft für Pflanzenkrankheiten*, 1934, **10**, 1059–1068.

48. P. F. Dowd, Fusaric acid a secondary fungal metabolites that synergizes toxicity of co-occurring host allelochemicals to corn earworm, *Heliothis zea* (Lepidoptera), *J. Chem. Ecol.*, 1989, **15**, 249–254.

49. T. K. Smith, E. G. McMillan and J. B. Castillo, Effect of feeding blends of *Fusarium* mycotoxin-contaminated grains containing deoxynivalenol and fusaric acid on growth and feed consumption of immature swine, *J. Anim. Sci.*, 1997, **75**, 2184–2191.

50. E. Gäumann, H. Kobel and S. Naef-Roth, *Über Fusarinsäure, ein zweites Welketoxin des Fusarium lycopersici Sacc*, Eidgenössische Technische Hoschule in Zürich, 1952.

51. E. Gaumann, Fusaric acid as a wilt toxin, *Phytopathology*, 1957, **47**, 342–357.

52. I. Gapillout, M.-L. Milat and J.-P. Blein, Effects of fusaric acid on cells from tomato cultivars resistant or susceptible to *Fusarium oxysporum* f. sp. *lycopersici*, *Eur. J. Plant Pathol.*, 1996, **102**, 127–132.

53. B. Bouizgarne, M. Brault, A. M. Pennarun, J. P. Rona, Y. Ouhdouch, I. El Hadrami and F. Bouteau, Electrophysiological responses to fusaric acid of root hairs from seedlings of date palm-susceptible and -resistant to *Fusarium oxysporum* f. sp. *Albedinis*, *J. Phytopathol.*, 2004, **152**, 321–324.

54. B. Bouizgarne, H. El-Maarouf-Bouteau, K. Madiona, B. Biligui, M. Monestiez, A. M. Pennarun, Z. Amiar, J. P. Rona, Y. Ouhdouch, I. El Hadrami and F. Bouteau, A Putative role for fusaric acid in biocontrol

of the parasitic angiosperm orobanche ramose-B, *Mol. Plant-Microbe Interact.*, 2006, **19**, 550–556.

55. B. Barna, C. Jansen and K.-H. Kogel, Sensitivity of barley leaves and roots to fusaric acid, but not to H_2O_2, is associated with susceptibility to *Fusarium* Infections, J. *Phytopathol.*, 2011, **159**, 720–725.

56. J. Horáček, L. Švábová, P. Šarhanová and A. Lebeda, Variability for resistance to *Fusarium solani* culture filtrate and fusaric acid among somaclones in pea, *Biol. Plant.*, 2013, **57**, 133–138.

57. E.-M. Niehaus, K. W. von Bargen, J. J. Espino, A. Pfannmueller, H.-U. Humpf and B. Tudzynski, Characterization of the fusaric acid gene cluster in *Fusarium fujikuroi*, *Appl. Microbiol. Biotechnol.*, 2014, **98**, 1749–1762.

58. D. Klüpfel, Über die Biosynthese und die Umwandlung der Fusarinsäure in Tomatenpflanzen, *Phytopathologische Zeitschrift*, 1957, **29**, 350–379.

59. J. D. Miller, J. C. Young and H. L. Trenholm, Decline in deoxynivalenol (vomitoxin) concentrations in 1983 Ontario winter wheat before harvest, *Can. J. Bot.*, 1983, **61**, 3080–3087.

60. J. Miller and P. Arnison, Degradation of deoxynivalenol by suspension culture of the Fusarium head blight resistant wheat cultivar Frontana, *Can. J. Plant Pathol.-Rev. Can. Phytopathol.*, 1986, **8**, 147–150.

61. N. Sewald, J. L. von Gleissenthall, M. Schuster, G. Müller and R. T. Aplin, Structure elucidation of a plant metabolite of 4-desoxynivalenol, *Tetrahedron: Asymmetry*, 1992, **3**, 953–960.

62. B. Poppenberger, F. Berthiller, D. Lucyshyn, T. Sieberer, R. Schuhmacher, R. Krska, K. Kuchler, J. Glossl, C. Luschnig and G. Adam, Detoxification of the *Fusarium* mycotoxin deoxynivalenol by a UDP-glucosyltransferase from *Arabidopsis thaliana*, *J. Biol. Chem.*, 2003, **278**, 47905–47914.

63. W. Schweiger, J. Boddu, S. Shin, B. Poppenberger, F. Berthiller, M. Lemmens, G. J. Muehlbauer and G. Adam, Validation of a candidate deoxynivalenol-inactivating UDP-glucosyltransferase from barley by heterologous expression in yeast, *Mol. Plant-Microbe Interact.*, 2010, **23**, 977–986.

64. P. Karlovsky, Biological detoxification of the mycotoxin deoxynivalenol and its use in genetically engineered crops and feed additives, *Appl. Microbiol. Biotechnol.*, 2011, **91**, 491–504.

65. G. Wiesenberger, E. Varga, C. Hametner, R. Stueckler, H. C. Kistler, T. J. Ward, D. Schoefbeck, R. Schuhmacher, F. Berthiller and G. Adam, North American isolates of *Fusarium graminearum* produce a novel type A trichothecene, *36th Mycotoxin Workshop*, 2014, Göttingen, Germany, p. 31.

66. K. Clay and C. Schardl, Evolutionary origins and ecological consequences of endophyte symbiosis with grasses, *Am. Nat.*, 2002, **160**(Suppl 4), S99–S127.

67. F. Posada and F. E. Vega, Establishment of the fungal entomopathogen *Beauveria bassiana* (Ascomycota: Hypocreales) as an endophyte in cocoa seedlings (*Theobroma cacao*), *Mycologia*, 2005, **97**, 1195–1200.
68. R. A. Campos, J. T. Boldo, I. C. Pimentel, V. Dalfovo, W. L. Araújo, J. L. Azevedo, M. H. Vainstein and N. M. Barros, Endophytic and entomopathogenic strains of *Beauveria* sp. to control the bovine tick Rhipicephalus (Boophilus) microplus, *Genet. Mol. Res.*, 2010, **9**, 1421–1430.
69. F. E. Vega, F. Posada, M. Catherine Aime, M. Pava-Ripoll, F. Infante and S. A. Rehner, Entomopathogenic fungal endophytes, *Biol. Control*, 2008, **46**, 72–82.
70. S. D. Reay, M. Brownbridge, B. Gicquel, N. J. Cummings and T. L. Nelson, Isolation and characterization of endophytic *Beauveria* spp. (Ascomycota: Hypocreales) from *Pinusradiata* in New Zealand forests, *Biol. Control*, 2010, **54**, 52–60.
71. A. G. Mcinnes, D. G. Smith, C.-K. Wat, L. C. Vining and J. L. C. Wright, , New techniques in biosynthetic studies using ^{13}C nuclear magnetic resonance spectroscopy. The biosynthesis of tenellin enriched from singly and doubly labelled precursors, *J. Chem. Soc., Chem. Commun.*, 1974, **8**, 281–282.
72. B. Schulz and C. Boyle, The endophytic continuum, *Mycol. Res.*, 2005, **109**, 661–686.
73. B. H. Ownley, M. R. Griffin, W. E. Klingeman, K. D. Gwinn, J. K. Moulton and R. M. Pereira, *Beauveria bassiana*: endophytic colonization and plant disease control, *J. Invertebr. Pathol.*, 2008, **98**, 267–270.
74. D. Jaeschke, D. Dugassa-Gobena, P. Karlovsky, S. Vidal and J. Ludwig-Mueller, Suppression of clubroot (*Plasmodiophora brassicae*) development in *Arabidopsis thaliana* by the endophytic fungus *Acremonium alternatum*, *Plant Pathol.*, 2010, **59**, 100–111.
75. G. A. Strobel and W. M. Hess, Glucosylation of the peptide leucinostatin A, produced by an endophytic fungus of European yew, may protect the host from leucinostatin toxicity, *Chem. Biol.*, 1997, **4**, 529–536.
76. A. Ishiyama, K. Otoguro, M. Iwatsuki, M. Namatame, A. Nishihara, K. Nonaka, Y. Kinoshita, Y. Takahashi, R. Masuma, K. Shiomi, H. Yamada and S. Ōmura, In vitro and in vivo antitrypanosomal activities of three peptide antibiotics: leucinostatin A and B, alamethicin I and tsushimycin, *J. Antibiot.*, 2009, **62**, 303–308.
77. S. El-Sharkawy and Y. J. Abul-Hajj, Microbial cleavage of zearalenone, *Xenobiotica*, 1988, **18**, 365–371.
78. M. Stob, R. S. Baldwin, J. Tuite, F. N. Andrews and K. G. Gillette, Isolation of an anabolic, uterotrophic compound from corn infected with *Gibberella zeae*, *Nature*, 1962, **196**, 1318.
79. C. M. Christensen, G. H. Nelson and C. J. Mirocha, Effect on the white rat uterus of a toxic substance isolated from *Fusarium*, *Appl. Microbiol.*, 1965, **13**, 653–659.

80. C. J. Mirocha, C. M. Christensen and G. H. Nelson, Estrogenic metabolite produced by *Fusarium graminearum* in stored corn, *Appl. Microbiol.*, 1967, **15**, 497–503.

81. G. Engelhardt, G. Zill, B. Wohner and P. R. Wallnöfer, Transformation of the *Fusarium* mycotoxin zearalenone in maize cell suspension cultures, *Naturwissenschaften*, 1988, **75**, 309–310.

82. P. M. P. Kovalsky, W. Schweiger, C. Hametner, R. Stückler, G. J. Muehlbauer, E. Varga, R. Krska, F. Berthiller and G. Adam, Zearalenone-16-O-glucoside: a new masked mycotoxin, *J. Agric. Food Chem.*, 2014, **62**, 1181–1189.

83. U. Matern and G. K. Tietjen, *Phytotoxins in Plant Pathogenesis*, ed. A. Graniti, R. D. Durbin and A. Ballio, NATO ASI Series, Ser. H27, Springer, Heidelberg, 1989, pp. 419–421.

84. R. E. Kneusel, U. Matern, V. Wray and K. D. Klöppel, Detoxification of the macrolide toxin brefeldin a by *Bacillus subtilis*, *FEBS Lett.*, 1990, **275**, 107–110.

85. H. Kakeya, N. Takahashi-Ando, M. Kimura, R. Onose, I. Yamaguchi and H. Osada, Biotransformation of the mycotoxin, zearalenone, to a non-estrogenic compound by a fungal strain of *Clonostachys* sp., *Biosci. Biotechnol. Biochem.*, 2002, **66**, 2723–2726.

86. N. Takahashi-Ando, M. Kimura, H. Kakeya, H. Osada and I. Yamaguchi, A novel lactonohydrolase responsible for the detoxification of zearalenone: enzyme purification and gene cloning, *Biochem. J.*, 2002, **365**, 1–6.

87. I. Matthies, G. Woerfel and P. Karlovsky, Induction of a zearalenone degrading enzyme caused by the substrate and its derivatives, *Mycotoxin Res*, 2001, **17**(Suppl 1), 28–31.

88. E. H. Crane III, J. T. Gilliam, P. Karlovsky and J. R. Maddox, patent application *Compositions and methods of zearalenone detoxification*, 2002, WO 2002/076205.

89. P. Karlovsky, E. H. Crane III, J. T. Gilliam and J. R. Maddox, patent application *Compositions and methods of zearalenone detoxification*, 2003, US 2003/0073239.

90. W. H. Urry, H. L. Wehrmeister, E. B. Hodge and P. H. Hidy, The structure of zearalenone, *Tetrahedron Lett.*, 1966, 7, 3109–3114.

91. C. Kosawang, M. Karlsson, H. Vélëz, P. H. Rasmussen, D. B. Collinge, B. Jensen and D. F. Jensen, Zearalenone detoxification by zearalenone hydrolase is important for the antagonistic ability of *Clonostachys Rosea* against mycotoxigenic *Fusarium graminearum*, *Fungal Biol.*, 2014, **118**, 364–373.

92. D. Popiel, G. Koczyk, A. Dawidziuk, K. Gromadzka, L. Blaszczyk and J. Chelkowski, Zearalenone lactonohydrolase activity in *Hypocreales* and its evolutionary relationships within the epoxide hydrolase subset of a/b-hydrolases, *BMC Microbiol.*, 2014, **14**, 82.

93. E. Vekiru, C. Hametner, R. Mitterbauer, J. Rechthaler, G. Adam, G. Schatzmayr, R. Krska and R. Schuhmacher, Cleavage of zearalenone

by *Trichosporon mycotoxinivorans* to a novel nonestrogenic metabolite, *Appl. Environ. Microbiol.*, 2010, **76**, 2353–2359.

94. R. Utsumi, T. Hadama, M. Noda, H. Toyoda, H. Hashimoto and S. Ohuchi, Cloning of fusaric acid-detoxifying gene from *Cladosporium werbeckii*: a new strategy for prevention of plant diseases, *J. Biotechnol.*, 1988, **8**, 311–316.

95. Y. Shibano, H. Toyoda, R. Utsumi and K. Obata, *Fusaric acid resistant genes*, 1991, patent application EP0444664.

96. Y. Shibano, H. Toyoda, R. Utsumi and K. Obata, *Fusaric acid resistant genes*, 1994, patent application US5292643.

97. H. Toyoda, H. Hashimoto, R. Utsumi, H. Kobayashi and S. Ouchi, Detoxification of fusaric-acid by a fusaric-acid-resistant mutant *of Pseudomonas solanacearum* and its application to biological control of *Fusarium* wilt of tomato, *Phytopathology*, 1988, **78**, 1307–1311.

98. H. Toyoda, K. Katsuragi, T. Tamai and S. Ouchi, DNA sequence of genes for detoxification of Fusaric acid, a wilt-inducing agent produced by *Fusarium* species, *J. Phytopathol.*, 1991, **133**, 265–277.

99. R. Utsumi, T. Yagi, S. Katayama, K. Katsuragi, K. Tachibana, H. Toyoda, S. Ouchi, K. Obata, Y. Shibano and M. Noda, Molecular cloning and characterization of the fusaric acid resistance gene from *Pseudomonas cepacia*, *Agric. Biol. Chem.*, 1991, **55**, 1913–1918.

100. K. Poole, K. Krebes, C. McNally and S. Neshat, Multiple antibiotic resistance in *Pseudomonas aeruginosa*: evidence for involvement of an efflux operon, *J. Bacteriol.*, 1993, **175**, 7363–7372.

101. J. L. Burns, C. D. Wadsworth, J. J. Barry and C. P. Goodall, Nucleotide sequence analysis of a gene from *Burkholderia (Pseudomonas) cepacia* encoding an outer membrane lipoprotein involved in multiple antibiotic resistance, *Antimicrob. Agents Chemother.*, 1996, **40**, 307–313.

102. W. Fakhouri, F. Walker, W. Armbruster and H. Buchenauer, Detoxification of fusaric acid by a nonpathogenic *Colletotrichum* sp. *Physiol, Mol. Plant Pathol.*, 2003, **63**, 263–269.

103. F. K. Crutcher, J. Liu, L. S. Puckhaber, R. D. Stipanovic, S. E. Duke, A. A. Bell, H. J. Williams and R. L. Nichols, Conversion of fusaric acid to fusarinol by *Aspergillus tubingensis*: a detoxification reaction, *J. Chem. Ecol.*, 2014, **40**, 84–89.

104. I. Yamaguchi, M. Kimura, N. Ando, A. Nishiyama, T. Fukuda, H. Kakeya and H. Osada, patent application *Zearalenone-detoxifying enzyme, gene and transformant having the gene transferred thereinto* (in Japanese), 2003, WO 2003/080842.

Concluding Remarks

CHIARA DALL'ASTA[*a] AND FRANZ BERTHILLER[b]

[a] Department of Food Science, University of Parma, Parco Area Scienze 17/A, 43124 Parma, Italy; [b] Center for Analytical Chemistry, Department for Agrobiotechnolgy (IFA-Tulln), University of Natural Resources and Life Sciences, Vienna (BOKU), Konrad Lorenz Straße 20, 3430 Tulln, Austria
*Email: chiara.dallasta@unipr.it

9.1 Masked Mycotoxins

The term "masked mycotoxin" was used for the first time in 1990, referring to a glycosidic derivative of zearalenone found in corn and cleaved to its parent compound in the gastrointestinal tract of pigs.[1] Studies progressed slowly until 2005, when deoxynivalenol-3-glucoside (DON-3-Glc) was isolated from wheat and fully characterized.[2] In the same period, the role played by the masking mechanism exerted by the plant in the *Fusarium* head blight (FHB) pathogenesis and resistance in wheat was defined as well.[3] From that moment on, a growing number of studies was published all over the world in an exponential trend. Today, 153 papers are listed in the ISI database when "masked mycotoxin" is used as the keyword (h-index: 23, Web of Science®, June 2015).

Initially, the definition of "masked mycotoxins" highlighted the difficulties in the detection of these compounds by routine analysis.[2] Today, the scenario has definitely changed. The wide diffusion of mass spectrometry (MS) instrumentation has allowed many groups to implement multi-analyte methods, making MS techniques the methods of choice in (masked) mycotoxin determination. High-resolution MS has greatly simplified the discovery of novel compounds. Although certified reference

Issues in Toxicology No. 24
Masked Mycotoxins in Food: Formation, Occurrence and Toxicological Relevance
Edited by Chiara Dall'Asta and Franz Berthiller

Published by the Royal Society of Chemistry, www.rsc.org

materials are still lacking for masked mycotoxins, there is an increased availability of analytical standards and pure compounds.

9.2   Future Perspectives

Over the years, these changes have led to a significant increase in occurrence data, both in grains and in products thereof. The data clearly indicate the ubiquitous presence of the masked forms of mycotoxins, in amounts of up to 30% of the parent form. In some cases, this percentage can increase and even double the amount of the parent toxin, especially in whole meal or fiber-enriched products. This opens up new challenges for both risk assessment and industrial production, given the worldwide tendency to promote the intake of fiber in a healthy diet. Following the growing interest of the scientific community and the increasing output of research, masked mycotoxins are now in the pipeline of European Union (EU) legislation. Recently, the EU has given a mandate to the European Food Safety Authority (EFSA) to evaluate the risk related to the occurrence of masked mycotoxins in food and feed.[4] The scientific opinion addressed the assessment of the co-occurrence of masked and modified forms in food and feed, together with their parent compounds. Accordingly, the risk associated with dietary exposure was evaluated for humans and animals. While a safe scenario for humans was obtained by applying a lower bound approach, some concern may arise for consumers of large amounts at the upper bound. As for animals, difficulties arose in risk estimation due to some deficits in toxicological data (no observed adverse effect level/lowest observed adverse effect level values). Most toxicological studies so far have been performed by adding pure mycotoxins to feed, instead of naturally incurred grains. Nonetheless, masked mycotoxins have been considered as of low concern for livestock.

In addition to defining the scientific bases of the next regulation, the EFSA opinion has identified some significant gaps in the research, mainly in terms of toxicological aspects. The need for validated methods of routine analysis is strongly warranted as well. From a toxicological perspective, masked mycotoxins are commonly considered toxicologically relevant on account of the possible release of the parent forms upon digestion. The cleavage of masked zearalenone at the gastrointestinal level was demonstrated in 1990 by Gareis *et al.*[1] Confirmation of the cleavage of DON-3-Glc was provided by two independent studies published simultaneously in 2014.[5,6] Both works considered the capacity of the human gut microflora to metabolize the masked forms, releasing DON in the first stage and further biotransforming this into as-yet unknown compounds. These results were further confirmed *in vivo* for DON-3-Glc in piglets by Nagl *et al.*[7]

From a complementary and antithetical position, the scientific debate is now focusing on the physiological and toxicological role of modified mycotoxins *per se* and not simply in consideration of their possible cleavage. As distinct compounds other than their parents, modified mycotoxins may enter different pathways or have different toxicological targets. Similarly, they can interact differently with detoxifying enzymes and/or bacteria. Very little is known so far, so this possibly different behavior must be investigated at both the macro level and the molecular level. The outcome of these toxicological studies will actually drive the legislators towards setting comprehensive regulation, covering both parent and modified mycotoxins. As a consequence, while stakeholders of the agro-food system will deal with new guidelines, the scientific community will be challenged with some gaps in the research, and its focus will be shifted away from this as a purely analytical issue.

Among the new targets for research studies, the understanding of the fate of modified compounds along the food/feed production chain and their roles in plants will be primary from an industrial perspective. A deeper knowledge about the transfer rate and the balance between parent and modified forms along the food chain will support more targeted and cost-effective auto-control measures and monitoring plans, while opening up innovative strategies of mitigation. Similarly, the safe and efficient use of microbial strains—or enzymes isolated thereof—for biological detoxification/biotransformation may represent both a challenge and an opportunity for food and feed safety in the forthcoming years. The identification and exploitation of "green" and sustainable biocontrol and/or biotransforming agents will open up new opportunities in the agro-food system, especially in terms of reducing crop losses and increasing food security.

Another hot aspect in the masked mycotoxin field is the understanding of resistance mechanisms in plants. As very recent studies clearly demonstrate that DON-3-Glc *per se* is far less toxic than DON,[8] breeders may exploit the biosynthetic pathway leading to masking in order to increase plant resistance to FHB. In this context, metabolomics may support the progress in functional genomics for the selection of resistant varieties. Indeed, resistance-related metabolites can be identified by metabolic profiling. Thus, such studies may effectively contribute to improving our knowledge about biosynthesis and the regulation of metabolic pathways in plants.[9] The understanding of plant–pathogen cross-talk could be considerably boosted by the use of -omics approaches, as they can speed up the interpretation of plant mechanisms of resistance to biotic and abiotic stress factors.[10]

From a toxicological point of view, the investigation of the role of gut microbiota in the biotransformation of modified mycotoxins is one of the most intriguing challenges. As often stated, the microbiota is the largest organ in a body, and it is strongly influenced not only by species-related aspects, but also by other factors, such as age, gender, health status and, significantly, diet. Recent metagenomic studies confirm that diet–microbe interactions in the gut are critical for human health and disease, clearly indicating the key role played by the microbiota in affecting the immune system and homeostasis. In addition to positive bioactive compounds such as polyphenols, the intestinal microbiota is able to biotransform and affect the uptake of xenobiotics. Similarly, xenobiotics may unbalance the microbiota. This mutual relationship should thus be carefully taken into account when the toxicological relevance of xenobiotics is considered.[11] As an example, non-specific biotransformation can be considered as ubiquitous, but other metabolic pathways may be active/inactive according to the specific characteristics of the host. For example, the aforementioned studies of masked mycotoxins suggested that the mechanism of DON-3-Glc deglycosylation appeared to be non-specific in the human gut.[5,6] By contrast, DON de-epoxidation to DOM-1, a widespread metabolite in animals, is significantly less prevalent in humans. Understanding the role of the intestinal microflora, as well as the re-uptake of released parent forms in the intestine and the possible toxicity of the biotransformation products, are thus fundamental aspects in the determination of the true relevance of masked mycotoxins on human health. The role played by the gut microbiota and the ability to modulate it through diet is therefore a new frontier of research not only for masked mycotoxins, but for xenobiotics in general.

After more than a decade of intensive research on masked mycotoxins, it is a commonly held feeling that the term "masked" has lost is primary meaning. However, its meaning can be easily extended beyond the capability to elude detection analysis. Starting from the need for intermolecular interaction for exerting any biological activity, any chemical modification able to reduce or prevent such interactions can be exploited for detoxification, with the final aim of mitigating adverse effects in living organisms—plants, animals and humans. Thus, the question as to whether the "mask" hides mycotoxins from biological targets actually raises much more interest and opens up new frontiers for the current and next generation of researchers.

References

1. M. Gareis, J. Bauer, J. Thiem, G. Plank, S. Grabley and B. Gedek, Cleavage of zearalenone-glycoside, a "masked" mycotoxin, during digestion in swine, *J. Vet. Med. Ser. B*, 1990, **37**, 236–240.
2. F. Berthiller, C. Dall'Asta, R. Schuhmacher, M. Lemmens, G. Adam and R. Krska, Masked Mycotoxins: Determination of a Deoxynivalenol Glucoside in Artificially and Naturally Contaminated Wheat by LC-MS/MS, *J. Agric. Food Chem.*, 2005, **53**, 3421–3425.
3. M. Lemmens, U. Scholz, F. Berthiller, A. Koutnik, C. Dall'Asta, R. Schuhmacher, G. Adam, A. Mesterhazy, R. Krska, H. Buerstmayr and P. Ruckenbauer, A major QTL for Fusarium head blight resistance in wheat is correlated with the ability to detoxify the mycotoxin deoxynivalenol, *Mol. Plant-Microbe Interact.*, 2005, **18**, 1318–1324.
4. EFSA Panel on Contaminants in the Food Chain, Scientific Opinion on the risks for human and animal health related to the presence of modified forms of certain mycotoxins in food and feed, *EFSA J*, 2014, **12**, 3916.
5. A. Dall'Erta, M. Cirlini, M. Dall'Asta, D. Del Rio, G. Galaverna and C. Dall'Asta, Masked mycotoxins are efficiently hydrolyzed by human

colonic microbiota releasing their aglycones, *Chem. Res. Toxicol.*, 2013, **26**, 305–312.

6. S. W. Gratz, G. Duncan and A. J. Richardson, The Human Fecal Microbiota Metabolizes Deoxynivalenol and Deoxynivalenol-3-Glucoside and May Be Responsible for Urinary Deepoxy-Deoxynivalenol, *Appl. Environ. Microbiol.*, 2013, **79**, 1821–1825.

7. V. Nagl, B. Woechtl, H. E. Schwartz-Zimmermann, I. Hennig-Pauka, W. D. Moll, G. Adam and F. Berthiller, Metabolism of the masked mycotoxin deoxynivalenol-3-glucoside in pigs, *Toxicol. Lett.*, 2014, **229**, 190–197.

8. A. Pierron, S. Mimoun, L. Murate, N. Loiseau, Y. Lippi, A. P. Bracarense, L. Liaubet, G. Schatzmayr, F. Berthiller, W. D. Moll and I. P. Oswald, Intestinal toxicity of the masked mycotoxin deoxynivalenol-3-β-D-glucoside, *Arch. Toxicol.*, 2015, DOI: 10.1007/s00204-015-1592-8.

9. B. Warth, A. Parich, C. Bueschl, D. Schoefbeck, N. K. Neumann, B. Kluger, K. Schuster, R. Krska, G. Adam, M. Lemmens and R. Schuhmacher, GC-MS based targeted metabolic profiling identifies changes in the wheat metabolome following deoxynivalenol treatment, *Metabolomics*, 2015, **11**, 722–738.

10. A. C. Kushalappa and R. Gunnaiah, Metabolo-proteomics to discover plant biotic stress resistance genes, *Trends Plant Sci.*, 2013, **18**, 522–531.

11. *Diet-Microbe Interactions in the Gut*, ed. K. Tuohy and D. Del Rio, Academic Press, Elsevier, London, UK, 2015.

Subject Index